T0203572

Smart Computing and Self-Adaptive Systems

Computational Intelligence Techniques
Series Editor: Vishal Jain

The objective of this series is to provide researchers a platform to present state-of-the-art innovations, research and design and implement methodological and algorithmic solutions to data processing problems, designing and analyzing evolving trends in health informatics and computer-aided diagnosis. This series provides support and aid to researchers involved in designing decision support systems that will permit societal acceptance of ambient intelligence. The overall goal of this series is to present the latest snapshot of ongoing research as well as to shed further light on future directions in this space. The series presents novel technical studies as well as position and vision papers comprising hypothetical/speculative scenarios. The book series seeks to compile all aspects of computational intelligence techniques from fundamental principles to current advanced concepts. For this series, we invite researchers, academicians and professionals to contribute, expressing their ideas and research in the application of intelligent techniques to the field of engineering in handbook, reference or monograph volumes.

Computational Intelligence Techniques and Their Applications to Software Engineering Problems
Ankita Bansal, Abha Jain, Sarika Jain, Vishal Jain, Ankur Choudhary

Smart Computational Intelligence in Biomedical and Health Informatics
Amit Kumar Manocha, Mandeep Singh, Shruti Jain, Vishal Jain

Data Driven Decision Making using Analytics
Parul Gandhi, Surbhi Bhatia and Kapal Dev

Smart Computing and Self-Adaptive Systems
Simar Preet Singh, Arun Solanki, Anju Sharma, Zdzislaw Polkowski and Rajesh Kumar

For more information about this series, please visit: https://www.routledge.com/Computational-Intelligence-Techniques/book-series/CIT

Smart Computing and Self-Adaptive Systems

Edited by
Simar Preet Singh
Arun Solanki
Anju Sharma
Zdzislaw Polkowski
Rajesh Kumar

CRC Press
Taylor & Francis Group
Boca Raton London New York

CRC Press is an imprint of the
Taylor & Francis Group, an **informa** business

First edition published 2022
by CRC Press
6000 Broken Sound Parkway NW, Suite 300, Boca Raton, FL 33487-2742

and by CRC Press
2 Park Square, Milton Park, Abingdon, Oxon, OX14 4RN

CRC Press is an imprint of Taylor & Francis Group, LLC

Library of Congress Cataloging-in-Publication Data
Names: Singh, Simar Preet, editor.
Title: Smart computing and self-adaptive systems / edited by Simar Preet
Singh, Arun Solanki, Anju Sharma, Zdzislaw Polkowski, and Rajesh Kumar.
Description: First edition. I Boca Raton, FL : CRC Press, 2022. I Series:
Computational intelligence techniques I Includes bibliographical
references and index.
Identifiers: LCCN 2021031418 (print) I LCCN 2021031419 (ebook) I ISBN
9780367741105 (hbk) I ISBN 9780367741129 (pbk) I ISBN 9781003156123 (ebk)
Subjects: LCSH: Expert systems (Computer science) I Adaptive computing
systems. I Artificial intelligence--Industrial applications.
Classification: LCC QA76.76.E95 S626 2022 (print) I LCC QA76.76.E95
(ebook) I DDC 006.3/3--dc23/eng/20211014
LC record available at https://lccn.loc.gov/2021031418
LC ebook record available at https://lccn.loc.gov/2021031419

ISBN: 978-0-367-74110-5 (hbk)
ISBN: 978-0-367-74112-9 (pbk)
ISBN: 978-1-003-15612-3 (ebk)

DOI: 10.1201/9781003156123

Typeset in Times
by SPi Technologies India Pvt Ltd (Straive)

Contents

Editors

Dr. Simar Preet Singh, presently an assistant professor in Computer Science and Engineering Department at School of Engineering and Applied Sciences, Bennett University, Greater Noida, Uttar Pradesh, is also a Microsoft Professional. His certifications include Microsoft Certified System Engineer (MCSE), Microsoft Certified Technology Specialist (MCTS) and Core Java. He had also undergone the training program for VB.Net and Cisco Certified Network Associates (CCNA). He has worked with Infosys Limited, DAV University, Jalandhar, Chandigarh Group of Colleges, Landran, Mohali, and GNA University, Phagwara. He also has International Trip and International Project to his name. He has published SCI/SCIE/Scopus-Indexed papers. He has presented many research papers in various National and International Conferences in India and abroad. In addition to these, he is an editor of multiple books and also has patents to his name. His areas of interests include Cloud Computing, Fog Computing, Internet of Things (IoT), Machine Learning and Bigdata.

Dr. Arun Solanki is working as an assistant professor in the Department of Computer Science and Engineering, Gautam Buddha University, Greater Noida, India, where he has been working since 2009. He has worked as Time Table Coordinator, member Examination, Admission, Sports Council, Digital Information Cell and other university teams from time to time. He has received M. Tech. degree in computer engineering from YMCA University, Faridabad, Haryana, India. He has received his Ph.D. in computer science and engineering from Gautam Buddha University in 2014. He has supervised more than 60 M. Tech. dissertations under his guidance.

His research interests span Expert System, Machine Learning and Search Engines. He has published many research articles in SCI/Scopus indexed International journals/conferences like IEEE, Elsevier, Springer, etc. He has participated in many international conferences, and has been a technical and advisory committee member of many conferences. He has organized several FDP, conferences, workshops and seminars. He has chaired many sessions at International Conferences. Dr. Solanki is working as an associate editor in *International Journal of Web-Based Learning and Teaching Technologies* (*IJWLTT*) IGI publisher. He has also been working as guest editor for special issues in *Recent Patents on Computer Science*, Bentham

Science Publishers. He is the editor of many books with a reputed publisher like IGI Global, CRC and AAP, and is working as the reviewer in Springer, IGI Global, Elsevier and other reputed publisher journals.

Dr. Anju Sharma is currently working with Punjab State Aeronautical Engineering College Patiala, a Constituent College of Maharaja Ranjit Singh Punjab Technical University, Bathinda. Her research interests include Smart Grid computing, Cloud computing, Internet of Things (IoT) and Fog Computing. She has varied numbers of publications in International Journals and Conferences of repute. She is Senior Member of IACSIT (Senior Member of International Association of Computer Science and Information Technology) and professional member of ACM India, IEEE. She is an active member (TCM and Reviewer) of varied conferences.

Dr. Zdzislaw Polkowski is adjunct professor of The Karkonosze State University of Applied Sciences in Jelenia Góra, Poland. Also he is professor of UJW at Faculty of Technical Sciences and Rector's Representative for International Cooperation and Erasmus+ Program at the Jan Wyzykowski University Polkowice. Since 2019, he is also adjunct professor in Department of Business Intelligence in Management, Wroclaw University of Economics and Business, Poland. Moreover, he is visiting professor in University of Pitesti, Romania, WSG University Bydgoszcz, Poland, and adjunct professor in Marwadi University, India. He was the former dean of the Technical Sciences Faculty during the period 2009–2012 at UZZM in Lubin. He holds a PhD degree in computer science and management from Wroclaw University of Technology, postgraduate degree in microcomputer systems in management from University of Economics in Wroclaw and postgraduate degree IT in education from Economics University in Katowice. He obtained his engineering degree in Computer Systems in Industry from Technical University of Zielona Gora. He has published more than 120 papers in journals, 25 conference proceedings, including more than 55 papers in journals indexed in the Web of Science, Scopus, IEEE.

Dr. Polkowski is co-editor of seven books which have been published in Springer and CRC Press, Taylor & Francis. He served as a member of Technical Program Committee in many international conferences, journals in Poland, India, China, Iran, Syria, United Kingdom, Romania, Turkey and Bulgaria. To date he has delivered over 60 invited talks at different international conferences across various countries. He is also the member of the Board of Studies and expert member of the doctoral research committee in many universities in India. He is the member of the editorial board of several journals and has served as a reviewer in a wide range of international journals. His area of interests includes IT in Business and IoT in Business and Education Technology. He has successfully completed a research project on developing the innovative methodology of teaching business informatics funded by the European Commission. He also owns an IT SME consultancy company in Polkowice and Lubin, Poland.

Rajesh Kumar is currently working as a professor in Computer Science and Engineering Department, Thapar Institute of Engineering and Technology, Patiala. He obtained his Ph.D. from IIT Roorkee in 1993. During the last 24 years of service in Thapar Institute of Engineering and Technology, he has served in various distinguished positions. He has more than 25 years of UG and PG teaching and research experience. He has over 110 publications in various International and National Journals with 1800+ citations and Google H index 23. He has so far guided 13 Ph.D. and 25 M.E./M.Sc. theses. His current research interests include FANETs, Machine Learning, IoT, and Cloud and Fog Computing.

Contributors

Ritu Aggarwal
Maharishi Markandeshwar Institute of
   Computer Technology & Business
   Manage Ambala
Haryana, India

Sujata S. Alegavi
University of Mumbai
Mumbai, Maharashtra, India

M. Anitha
P.B. College of Engineering
Department of CSE
Chennai, Tamil Nadu, India

S.V. Annlin Jeba
Sree Buddha College of Engineering

Robina Gujral Bagga
Department of Electronics and
   Communication
Ambalika Institute of Management and
   Technology
Uttar Pradesh, India

Vinayak Ashok Bharadi
University of Mumbai
Ratnagiri, Mumbai, Maharashtra, India

D. K. Chaturvedi
Faculty of Engineering
Dayalbagh Educational Institute
   (deemed to be Univ.)
Dayalbagh, Agra, India

J. Dhilipan
Department of Computer Applications
SRM IST
Ramapuram Campus
Tamilnadu, India

S. Dorababu
Koneru Lakshmaiah Education Foundation
Green Fields
Vaddeswaram, Guntur, Kerala, India

S. Hrushikesava Raju
Koneru Lakshmaiah Education
   Foundation
Green Fields
Vaddeswaram, Guntur, Kerala, India

Ancy Jose
Sree Buddha College of Engineering

Beulah Joslyn Jose
Temple University

Deepinder Kaur
Department of Computer Science
   and Engineering
Chandigarh University
Gharuan, Punjab, India

Gaganpreet Kaur
Chandigarh University
Gharuan, Punjab, India

Gurdip Kaur
University of New Brunswick
New Brunswick, Canada

P. M. Kavitha
Department of Computer Science and
   Engineering
and
Department of Computer Applications
SRM Institute of Science and
   Technology
Chennai, Tamil Nadu, India

S. Kavitha
Koneru Lakshmaiah Education
   Foundation
Green Fields
Vaddeswaram, Guntur, Kerala, India

Sandeep Kumar
Gautam Buddha University
Greater Noida, India

V. Lakshmi Lalitha
Koneru Lakshmaiah Education Foundation
Green Fields
Vaddeswaram
Guntur, Kerala, India

Alok Mishra
Department of Applied Physics
Ambalika Institute of Management and
    Technology
Uttar Pradesh, India

B. Muruganantham
Department of Computer Science and
    Engineering
SRM Institute of Science and
    Technology
Chennai, Tamil Nadu, India

Zdzislaw Polkowski
Jan WyzykowskiUniversity
Polkowicw, Poland

Atul Prakash Prajapati
Faculty of Engineering
Dayalbagh Educational Institute
    (deemed to be Univ.)
Agra, Uttar Pradesh, India

Lakshmi Ramani Burra
PVP Siddhartha Institute of Technology
Vijayawada, Andhra Pradesh, India

Agusthiyar Ramu
Department of Computer Application
    (BCA)
SRM IST Ramapuram Campus
Tamil Nadu, Chennai

C. Rohan
Gautam Buddha University
Greater Noida
India

J. Shyamala Devi
SRM IST Ramapuram Campus
Chennai

Simar Preet Singh
Bennett University
Greater Noida, Uttar Pradesh, India

S. Sindhu
SRM IST
Ramapuram Campus

Arun Solanki
Gautam Buddha University
Greater Noida, Uttar Pradesh, India

N. Sunanda
Koneru Lakshmaiah Education Foundation
Vaddeswaram
Guntur, Kerala, India

S. Suriya
Department of Computer Application
    (BCA)
SRM IST Ramapuram Campus
Chennai, Tamil Nadu, India

Akash Tayal
Indira Gandhi Delhi Technical
    University for Women
Delhi, India

Kumar Anubhav Tiwari
Department of Ultrasound Research
    Institute
Kaunas University of Technology
Kaunas, Lithuania

Praveen Tumuluru
Koneru Lakshmaiah Education
    Foundation
Vaddeswaram
Guntur, Kerala, India

N. Vijayalakshmi
Department of Computer Applications
SRM IST
Ramapuram Campus
Tamilnadu, India

Saiyed Faiayaz Waris
Vignan's Foundation for Science
    Technology and Research
Vadlamudi, Guntur, Kerala, India

1 Using Luong and Bahdanau Attention Mechanism on the Long Short-Term Memory Networks

A COVID-19 Impact Prediction Case Study

Vinayak Ashok Bharadi
University of Mumbai, Ratnagiri, India

Sujata S. Alegavi
University of Mumbai, Mumbai, India

CONTENTS

DOI: 10.1201/9781003156123-1

1

1.1 INTRODUCTION: DEEP LEARNING

Deep Neural Networks (DNNs) have significantly impacted the field of machine learning and computer vision, specifically with the initiation of innovative deeper architectures like the residual and Convolutional Neural Networks (CNNs). DNNs are ubiquitously used for text, image audio as well as video-based data. The DNNs have exhibited a state-of-the-art performance for these applications [1–3].

A DNN, as compared to an artificial neural network, has many hidden layers with more weights and biases associated with each hidden layer [4], which improves the overall performance of approximating more complex functions.

In deep learning, generally, an input 'X' is taken and is used to predict an output 'Y'. Deep neural nets will reduce the prediction error by learning the patterns for the given inputs and outputs and applying this knowledge to the inputs that the model has never seen before (Figure 1.1).

In this network, the input layer has seven nodes for the numerical representation of data; the hidden layer has multiple nodes that take care of maximum computation whereas the output layer is generally one node that predicts the output. At every node of hidden layer the computation takes place based on two important components known as weight and bias (Figure 1.2).

As we can see, three inputs are coming from the input layer to the nodes of the first hidden layer. Depending on the weights of the input and the associated bias parameter, the learning in the network takes place [4]. This is known as the activation function. After the inputs are passed through all the nodes of the hidden layers, the output is predicted and the network evaluates the performance of the prediction by computing the loss function (also known as the cost function in regression type of problems) [5]. Various loss functions used in the literature, but the mean squared error (MSE) is the loss function. The deviation among the actual output and the prediction is evaluated. The difference between actual output and predicted output is back propagated to the network using different algorithms (gradient descent is one such common algorithm for weigh update). This process also modifies the biases in each node such that the loss function is reduced further [6, 7].

Deep learning is a complex field that handles different tasks like using CNNs for images and recurrent neural networks (RNNs) for NLP type of problems.[8], etc. Deep learning networks can be focused further on the features that are required for the use of attention networks. Attention networks have paved their way for deep learning [8].

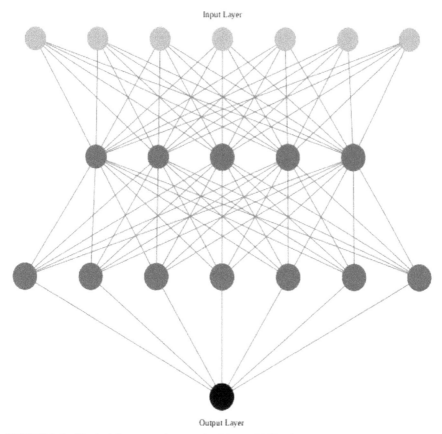

Input Layer

Output Layer

FIGURE 1.1 Typical deep neural network with two hidden layers.

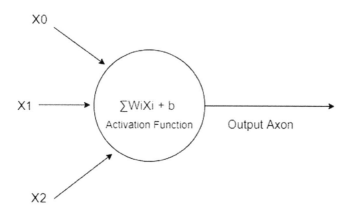

X0

X1 ⟶ ∑WiXi + b

Activation Function   Output Axon

X2

FIGURE 1.2 Node in a hidden layer.

1.2   RNN AND LSTMs

In recent years, due to the increase in the machines' computational power, deep learning techniques have gained a lot of popularity. Deep learning is a subcategory of neural networks that deals with training of input data on many layers with many numbers of hyperparameters. Depending on the input data, different architectures are used, such as DNNs, convolutional networks, and RNNs [9].

1.2.1   RECURRENT NEURAL NETWORKS

Generally, RNNs are used to deal with sequential data as DNN and CNN cannot deal with the temporal information present in the data. Two types of RNNs are generally used: (1) discrete-time RNN and (2) continuous-time RNN. RNN architecture has a recurring link to predict the future state built on current and past states as well as the current input data [9]. Generally, RNNs can be described as feed-forward neural networks with the capability of passing information across time steps. When input and/or output consists of sequences that are not independent, RNN is used [10].

In Figure 1.3, a simple RNN architecture is shown. The input layer is made up of two neurons that are connected to the hidden layers with hyperparameter settings. The output is predicted depending on this hidden layer and the past input received from the feedback loop between hidden layer 1 and hidden layer 2. To understand the sequential dependencies RNNs can be used over artificial neural networks. Due to the recurrent structure of the network, the memory that is being produced during the training phase is limited. Thus, all the algorithms used in the training phase have fallen victim to either vanishing or exploding gradients, resulting in the model's failure. The model won't learn the dependencies in the long-term sequential data. To address the issue of vanishing and the exploding gradients, Long-Short Term Memory (LSTM), popularly known as LSTM has been designed [11].

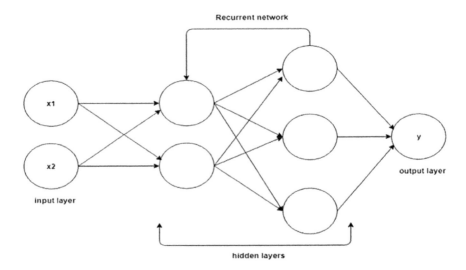

FIGURE 1.3   Simple RNN architecture.

1.2.2   LSTM ARCHITECTURE

LSTM architecture has matured over time. Many researchers have tried to change the basic architecture of the LSTM over time to suit various applications [12–14]. LSTM with only input and output gate and without a forget gate was used for irregularity detection in network traffic. Further, LSTM networks with forget gate and LSTM network having a peephole connection were also designed for various applications [14]. In this research, the LSTMs are used to predict sequence data; the COVID-19 case count is taken for the prediction study (Figure 1.4).

The LSTM architecture has three cell states known as gates. The input, forget, and the output gate [12, 15].

1.2.2.1   Input Gate

The input gate works for the update of the current cell state. The preceding hidden state and the latest input state are passed into the sigmoid function in the next step. The sigmoid function then decides which value will be updated by 0 and 1. Here, '0' indicates not important, whereas '1' indicates important. The tanh function is input with the hidden state and current inputs to adjust the values range between ±1 to control the network. The multiplication of tanh and sigmoid output is taken, and the sigmoid o/p decides the value from tanh output to be retained [16].

1.2.2.2   Forget Gate

The forget gate plays a major role in deciding whether to keep the information or to throw it away. If the forget gate value 'ft' becomes '1' then the cell state retains the information content, and if the value of the forget gate becomes '0' then the

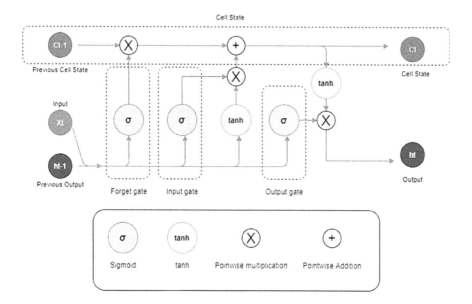

FIGURE 1.4   Conceptual diagram for long short-term memory network cell.

information is discarded. The bias bf attached with the forget gate can be increased, which in turn increases the performance of the LSTM network [17, 18].

1.2.2.3 Cell State

Forget vector and current state are multiplied pointwise to evaluate the cell state. The values in the cell state, which are multiplied by 0 are dropped. The final state is decided by the pointwise addition of the output of the input gate. Thus, the cell state is updated with relevant values only [19].

1.2.2.4 Output Gate

Next hidden state is generated by output gate. This hidden state retains the information about previous hidden state and has its application in the predictions. The sigmoid function receives the last hidden state and the latest input. The latest adapted cell state is fed to the tanh function. Output of tanh and sigmoid select the information to be retained, and hence the result is the hidden state. The new cell and hidden state are carried to the next time step [20].

The LSTM architecture, as discussed above, enables it to remember the relevant information for a longer duration and make use of the same for the prediction purpose. This solves the problem of vanishing gradients as observed in the RNNs [21].

1.3   SEQUENCE PREDICTION

Sequence prediction tasks are gaining much popularity as a special application of machine learning. The process of prediction of next output based on the previous input is known as sequence prediction. Sequence prediction can be used in various applications like predicting the next word in a series or predicting the next number, an event, an alphabet, or even an object. Before understanding the term sequence prediction, let us see what is considered a sequence in the applied machine learning field when observing certain data and recording it, forming a sample [22, 23]. Collection of such samples forms sets of observations that can be used in further processing. But, in such sets the order of observation is not considered, whereas in the case of sequence, it levies an obvious order on the findings. This order forms the base of prediction in sequence. Sequence prediction is a common problem which has real-time applications [24].

In sequence prediction, we can discuss many cases, but out of the three cases find applications in a lot of real-life problems. The first case deals with the prediction of the next value. Predicting the next value in a sequence is a very important machine learning task used in time series forecasting. A sequence prediction model is developed, which will learn from the previous input values and based on it will forecast the next value [25] (Figure 1.5).

The best example of sequence prediction of the next value is time series forecasting, in which there is an ordered series of data, where the observations are made sequentially in the time domain. The prediction of the next value that comes in the series is time series forecasting. Training the model on historical data in the past and using the knowledge for the future output prediction is denoted as time series

FIGURE 1.5   Value prediction.

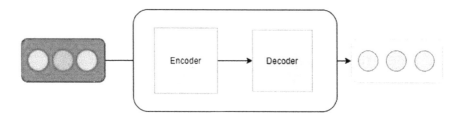

FIGURE 1.6   Class label prediction.

forecasting. The second case deals with the prediction of class labels. Unlike the prediction of values in sequence prediction, the model labelled datasets with some sequence inputs and output class labels taken by the model to predict the class labels for unseen sequences [26, 27] (Figure 1.6).

A wide area of applications linked with sequence classification includes health information systems, abnormality detection, and information retrieval. Sequence classification is challenging compared to normal classification tasks. By properly selecting feature vectors, the classification may be done easily in the case of image classification. But, in the case of sequence classification, as explicit features are unavailable even with highly complex classification models, the number of features may be huge, which causes dimensionality problems. There are three types of sequence classification models: the first category is a sequence classification model based on features. The feature vector is generated from the input sequence in a feature-based classification model and then passed to the conventional classifiers. The second group deals with the sequence distance-based classification. Model-based classification using Hidden Markov Models (HMMs) and other models of statistical nature for sequence classification forms the third group [25, 28] (Figure 1.7).

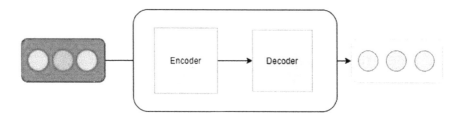

FIGURE 1.7   Sequence-to-sequence prediction.

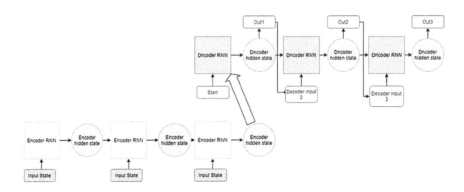

FIGURE 1.8   Overview of the sequence-to-sequence model.

In the seq2seq model, a word sequence is taken as input and an output sequence is generated. Generally, RNN is used in such cases, but as RNNs face hurdles due to vanishing or exploding gradients, advanced versions of RNNs like LSTM or Gated Recurrent Unit (GRU) are used for such applications. The word context is developed by considering two inputs: one from the previous step and one from the user. An encoder and a decoder section make the core of a seq2seq. The encoder uses deep learning network layers to generate hidden vectors corresponding to the input word sequences. The decoder takes its hidden states generated by the encoder and current word to generate the next hidden vector in view of the prediction of the next word [29].

Generally, the seq2seq models as shown in Figure 1.8 are optimized using certain optimization algorithms like attention mechanism, Beam search and Bucketing. Among all attention mechanisms attention is a very widely accepted optimization technique [30]. The decoder is fed with a single vector input that stores all the background context information, which results in bigger trouble when large sequences are used. In such cases, attention mechanism allows the decoders to optimize the algorithm by selective analysis of the input sequence [31].

In the convention seq2seq model, the last hidden state of the decoder RNN becomes the context vector. Further, the output is not accurate in the case of a long input sequence.

The attention mechanism directly addresses this issue by retaining and utilizing all the past hidden states of the input sequence as the decoding process progresses. An exclusive linkage between each time step of the decoder output and the hidden steps of the encoder is created to complete this task. Due to this mechanism, the decoder has access to the full input sequence and the decoder block can select specific elements from the input to generate the output.

1.4   ATTENTION MECHANISM

Neural networks were developed to imitate human brain functions more simply. When the brain needs to decide on a certain outcome, it defines certain parameters with different weights assigned to these parameters. The brain pays more attention to that parameter which it feels to be the most important. The neural network imitates the same action, wherein it selectively concentrates on a few important things and pays less importance to others; this technique is called the attention mechanism [32]. For example, if you are training a DNN with four parameters as your input weights and amongst the four parameters, the decision is majorly based on the first and second parameter; in such cases, the attention mechanism will focus its attention on parameters 1 and 2 ignoring 3 and 4. This will not only help to ease out the computational complexity of the network but also give better predictions. Previously LSTMs were designed in the encoder-decoder mode for various applications such as Natural Language Processing (NLP). The Neural Machine Translation (NMT) was implemented using the encoder-decoder LSTM units. The encoder is responsible for generating the context vector by processing the input. The final encoder state forms the initial state for the decoder ignoring all the other intermediate states. The context vector is processed using the decoder LSTM units, which produce the words in a sentence one after the other [33].

It is observed that the encoder makes sense of the entire sequence and summarizes it and the decoder simply decodes the sequence using the context vector. Here comes the problem, if the encoder badly summarizes the sequence, the context vector will be bad enough, and thus the decoder will output a bad sequence. This is seen in long-enough sentences where the encoder tends to create a bad summary, which is referred to as the long-range dependency problem of RNNs. Because of the vanishing/exploding gradient problems, RNNs cannot memorize long sequences. This problem is somewhat taken care of in LSTMs, which can capture the long-range dependencies, but, in some cases, it tends to become forgetful. While translating the sequence, there is no provision in LSTM to give importance to specific input words.

RNN's performance is observed to be degraded on longer input or output sequences. The use of fixed-sized internal representation by the encoder causes this issue. The attention mechanism supplements the architecture that addresses this problem. Attention provides a focused context from the encoder to the decoder. In the next step, a learning mechanism is also provided, enabling the decoder to specifically focus on the richer encoding for the prediction of each time step in the output sequence.

1.4.1 BAHDANAU ATTENTION MECHANISM

The NMT is focused on constructing an ANN adjusted to optimize translation perfor-mance [34]. In 2015, Bahdanau came up with a brilliant idea of considering all the input words while forming the context vector, but each input word is given relative importance. The advantage of this model is that it will search for all the hidden states in the encoder for the availability of the most relevant information [35]. In general terms, attention is one component in the architecture of the ANN; this mechanism critically impacts in quantifying and managing the linkage of the elements at the input and output end. This is referred to as General Attention. Another linkage within the input elements is referred to as the Self-Attention (Figure 1.9).

Let's start with an example where a source sequence x of length n is given, and it is required to output a target sequence y of length m:

$$x = \left[x_1, x_2, x_3, x_4, \ldots, x_n \right] \quad \text{and} \quad y = \left[y_1, y_2, y_3, y_4, \ldots y_m \right]$$

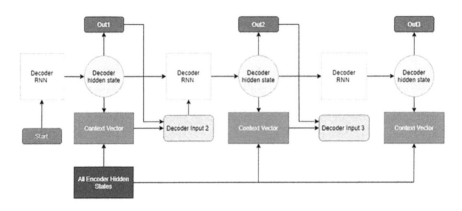

FIGURE 1.9   The Bahdanau attention model.

Annotation sequence (h_1, h_2, h_3, h_4, ...) is generated for every input sequence in the bidirectional LSTM. The onward and reverse hidden states in the encoder are concatenated; the output is used to generate the vectors h_1, h_2, ...

$$h_i = \left[\vec{h}_i^T ; \overleftarrow{h}_i^T \right]^T, \quad i = 1,\ldots,n \tag{1.1}$$

The Tx count of words as part of the input sequence is represented by all the vectors h_1, h_2, ... unlike the encoder-decoder model, the context vector is framed from the last state of the encoder only. This is done by evaluating the hidden state's weighted sum. Bahdanau suggests creating the context vector of all the words in the input sequence, where the feed-forward neural network calculates the weights. The context vector ci for the output word y_i is created by the weighted sum of the annotated inputs (Figure 1.10).

$$C_t = \sum_{i=1}^{n} \alpha_{t,i}^i h \tag{1.2}$$

The *SoftMax* function is used to calculate the weights α_{ij} given by the following equation: $\alpha_{i,j} = align(y_i, x_i)$ (How well two words y_i and xi are aligned)

$$\alpha_{ij} = \frac{\exp(e_{ij})}{\sum_{i=1}^{Tx} \exp(e_{ik})} \tag{1.3}$$

$$e_{ij} = a(s_{i-1}, h_j) \tag{1.4}$$

The alignment function captures the alignment of 'j' input and 'I' output. Considering the last decoder state having 'd' dimensions, 'Tx' number of hidden state vectors are also referred to as the encoder produce the annotations. They have

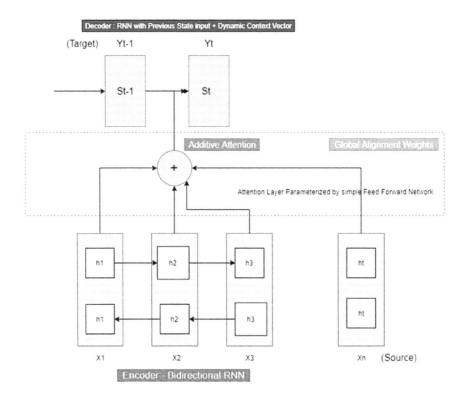

FIGURE 1.10 Additive attention implemented in encoder-decoder model as proposed by Bahdanau attention mechanism.

the same dimensions 'd'. Finally, the input length for the feed-forward network is (Tx, 2d). To get the scores e_{ij} having a dimension (Tx, 1), this input vector undergoes multiplication operation with matrix 'Wa' of (2d, 1) dimensions.

A *tanh* parabolic function is applied to these e_{ij} scores, which is finally given to the *SoftMax* function to get scores for the output. The SoftMax function results in enforcing the vector constituents to add up to 1, and every distinct value of the element will be in the range of 0 to 1; in effect, this represents input weightage at that time instance,

$$E = I \big[Tx * 2d \big] * Wa \big[2d * 1 \big] + B \big[Tx * 1 \big] \tag{1.5}$$

$$\alpha = \text{SoftMax} \big(\tanh (E) \big) \tag{1.6}$$

$$C = IT * \alpha \tag{1.7}$$

where α is a vector having (Tx, 1) dimensionsand consists of the weights matching to every word of the input sequence. Thus, we can conclude that Bahdanau provides a global attention approach that offers good results when most of the words in the sequences are important. Bahdanau attention is a type of global, soft as well as additive attention mechanism. In summary, the Bahdanau attention mechanism has the following steps:

1. Generation of the Encoder Hidden States
2. Calculation of the Alignment Scores
3. SoftMax of the Alignment Scores
4. Context Vector Generation
5. Output Decoding
6. Iteration: The procedure (as mentioned in steps 2–5) iterates until the target output of the desired length is generated.

1.5   LUONG ATTENTION MECHANISM

There is one problem with the global attention mechanism that for each target word it has to go through all the words from the input, which becomes quite expensive and impossible for translating longer sequences. To overcome this problem, Luong proposed a local attention mechanism, in which a minor subset of source positions for each target word is focused upon [36]. To reduce computation overhead during the soft attention mechanism and ease the training of hard attention, Luong's local attention mechanism considers a comparatively smaller window of separately identifiable context.

Multiplicative attention is another reference to Luong's attention. This attention mechanism was built over the Bahdanau attention as discussed earlier. Important points of difference are as follows:

1. The procedure for the calculation of the alignment scores
2. The location of the decoder time instance where the attention mechanism is being added.

In Luong's attention, initially, the aligned position p_t corresponding to every target word at time t is generated. First, the hidden source states in the context the window $[pt - D, pt + D]$ is taken, then weighted averaging over it is done. In this process, the D is an empirically selected term to derive the context vector c_t. In the case of Luong's, the fixed dimension alignment vector is generated; in the Bahdanau attention mechanism, a global context vector is observed (Figure 1.11).

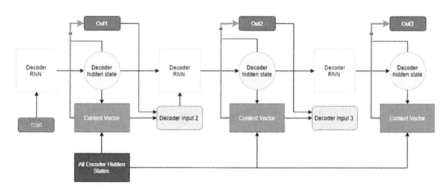

FIGURE 1.11   The Luong attention model.

The top LSTM layers are generated by the encoder and the decoder states, which are hidden. The bidirectional encoder concatenates the onward and reverse states which are hidden. Further, Luong uses ht → at → ct → ht to make a prediction. All three different scoring functions for alignment are used in Luong's attention model [36]. Bahdanau's attention mechanism has only one scoring fiction defined.

The architecture of the Attention Decoder for Luong attention mechanism is also different. In the case of Luong's attention, when the RNN generates the output for the time step under consideration, the context vector is consumed. In summary, Luong's attention mechanism has the following steps:

1. Generation of the Encoder Hidden States
2. RNN Decoder
3. Calculation of the Alignment Scores
4. SoftMax of the Alignment Scores
5. Context Vector Generation
6. Final output generation
7. Iteration: The procedure (as mentioned in steps 2–6) iterates every time instance of the decoder. This is done till the target output of the desired length is generated.

1.5.1 Comparison of Luong and Bahdanau Attention Mechanism

The topmost hidden layer states in both the encoder and the decoder are used in **Luong attention**. In the case of the **Bahdanau attention,** the concatenation Top Hidden Layer's forward and reverse hidden state is taken. In **Luong's attention,** they get the decoder's hidden state at time **t**. Then calculate attention scores and get the context vector that will be concatenated with the hidden state of the decoder and then predict. Still, in the **Bahdanau** at time **t** we consider about **t – 1** hidden state of the decoder.

In conclusion, the current decoder's hidden state is used in the **Luong** attention mechanism to compute the alignment vector. In contrast, the output of the previous time step is used in **Bahdanau** attention mechanism. Then the alignment is calculated and context vectors as discussed above. Then this context and the hidden state of the decoder are concatenated at **t – 1**. So, before the SoftMax, this concatenated vector goes inside a GRU. **Luong** has different types of alignments. **Bahdanau** has only a concatenated score alignment model. **Luong** is multiplicative attention; however, **Bahdanau** is an additive attention mechanism. **Luong** results in local attention along with global attention. The local attention is an amalgamation of hard and soft attention. With the proper design and implementation of the attention mechanism, powerful deep NLP and sequence prediction algorithms can be built.

1.6 COVID-19 DATA PREDICTION

The Novel Coronavirus has been a widespread pandemic of the 21st century [37]. Today, a total of 10.9 million cases have been reported, with 155K deaths. Since December 29, 2020 this virus has spread at an alarming rate and infected millions. The COVID-19 infection statistics given by John Hopkin's University [38] are used in this work. The coronavirus infection spread modelling is studied using the following ANN models:

1. Recurrent Neural Network
2. Simple seq2seq LSTM Model
3. Seq2Seq LSTM Model with Luong Attention
4. Encoder-Decoder with Bahdanau Attention

The prediction is compared with the actual figures.

The input data has the following heads for 115 days (till July 28, 2020)

1. Global Confirmed Cases
2. Global Deaths
3. Global Recovery
4. Confirmed Cases in India
5. Deaths in India
6. Recovery in India
7. Confirmed Cases in the USA
8. Deaths in the USA
9. Recovery in the USA

The plot for the data points can be seen in Figure 1.12.

The sequences mentioned above are studied in univariate as well as multivariate analysis. The LSTM is trained with these sequences for the prediction of the next values for these heads. This is programmed in python language using Keras API [39].

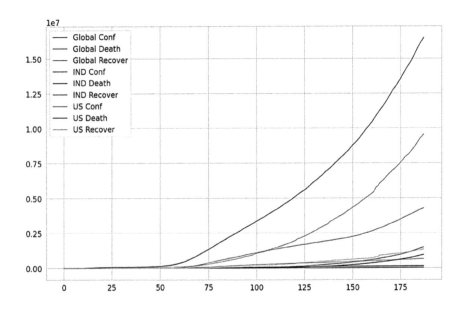

FIGURE 1.12   COVID-19 data (scaling factor—100000).

Francois Chollet a Developer/Researcher at Google AI created and developed the Keras framework [40]. Keras as default backend as Theano till the release of V1.1.0. Keras support TensorFlow from V1.1.0 onwards. The Keras v2.3reseased on September 17, 2019; this release of Keras was the first one that was in sync with tf [41].

1.7 LSTM IMPLEMENTATION USING KERAS

The deep learning approaches like LSTM sequential models, especially the seq2seq type move, displayed significant success in addressing time series problems by considering the multiple outputs and the long-term dependencies [42]. The LSTM-seq2seq model shows remarkable prediction ability and helps in improving the forecast accuracy in the short-term COVID-19 impact prediction case study.

The input data has the format as [Sample, Time Step, Features], there is a total of 148 training samples, each is presented as a single time step, and there are nine features as discussed above and shown in Figure 1.12. Kera's sequential model is used to build the encoder-decoder network.

The encoder-decoder model for LSTMs (RNNs) is designed for sequence-to-sequence (seq2seq) prediction problems. This architecture has two sub-models:

- **Encoder**: The encoder visits all the time steps and encodes the whole sequence. This is represented by a fixed-length called the context vector.
- **Decoder**: The decoder does the task of analysing the output time steps, parallelly scanning the context vector generated by the encoder.

All the models discussed below have been designed using encoder-decoder architecture for sequence prediction.

1.7.1 LSTM SEQUENCE-TO-SEQUENCE MODEL

The model summary for the sequential seq2seq is shown in Figure 1.13. First, the encoder is built, then the batch normalization layer is added, and later, the decoder is stacked. Batch normalization is added to avoid gradient explosion resulting from the ELU activation function in the encoder.

1.7.2 LSTM SEQUENCE-TO-SEQUENCE IMPLEMENTATION WITH LUONG ATTENTION

This implementation has slight variation, besides returning the last hidden state and the last cell state, it is also needed to return the stacked hidden states for alignment score calculation. The attention layer building process requires to calculate the alignment score and apply the SoftMax activation function on the same as follows:

```
attention   =   dot([decoder_stack_h,   encoder_stack_h],
   axes=[2, 2])
attention = Activation('softmax')(attention)
```

Then the context vector is calculated, and the batch normalization is applied on top of it:

```
context = dot([attention, encoder_stack_h], axes=[2,1])
context = BatchNormalization(momentum=0.6)(context)
```

```
Model: "functional_23"
_____
Layer (type)                    Output Shape          Param #    Connected to
===============================================================================
input_43 (InputLayer)           [(None, 1, 9)]         0
_____
lstm_15 (LSTM)                  [(None, 200), (None,   168000     input_43[0][0]
_____
batch_normalization (BatchNorma (None, 200)            800        lstm_15[0][0]
_____
repeat_vector (RepeatVector)    (None, 1, 200)         0          batch_normalization[0][0]
_____
batch_normalization_1 (BatchNor (None, 200)            800        lstm_15[0][2]
_____
lstm_16 (LSTM)                  (None, 1, 200)         320800     repeat_vector[0][0]
                                                                  batch_normalization[0][0]
                                                                  batch_normalization_1[0][0]
_____
time_distributed_7 (TimeDistrib (None, 1, 9)           1809       lstm_16[0][0]
===============================================================================
Total params: 492,209
Trainable params: 491,409
Non-trainable params: 800
```

(a)

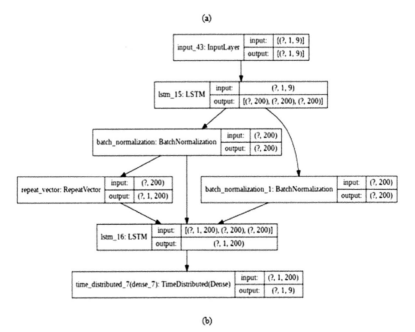

(b)

FIGURE 1.13 LSTM sequence-to-sequence (seq2seq) model: (a) Summary; (b) Model plot.

The context vector and stacked hidden states of decoder are joined together and use last dense layer input

```
decoder_combined_context = concatenate([context, decoder_
stack_h])
```

The final model with Luong attention is as follows (Figure 1.14):

As discussed earlier, the batch normalization is also added. Figure 1.15 shows the plot for the LSTM model with Luong attention. The model with and without batch normalization is implemented.

```
Model: "functional_25"
_____
Layer (type)                  Output Shape          Param #   Connected to
=================================================================================
input_45 (InputLayer)         [(None, 1, 9)]         0
_____
lstm_18 (LSTM)                [(None, 1, 200), (No   168000    input_45[0][0]
_____
repeat_vector_1 (Repeatvector) (None, 1, 200)        0         lstm_18[0][1]
_____
lstm_19 (LSTM)                (None, 1, 200)         320800    repeat_vector_1[0][0]
                                                               lstm_18[0][1]
                                                               lstm_18[0][2]
_____
dot (Dot)                     (None, 1, 1)           0         lstm_19[0][0]
                                                               lstm_18[0][0]
_____
activation (Activation)       (None, 1, 1)           0         dot[0][0]
_____
dot_1 (Dot)                   (None, 1, 200)         0         activation[0][0]
                                                               lstm_18[0][0]
_____
batch_normalization_2 (BatchNor (None, 1, 200)       800       dot_1[0][0]
_____
concatenate (Concatenate)     (None, 1, 400)         0         batch_normalization_2[0][0]
                                                               lstm_19[0][0]
_____
time_distributed_8 (TimeDistrib (None, 1, 9)         3609      concatenate[0][0]
=================================================================================
Total params: 493,209
Trainable params: 492,809
Non-trainable params: 400
```

FIGURE 1.14 Summary for the LSTM sequence-to-sequence (seq2seq) model with Luong attention.

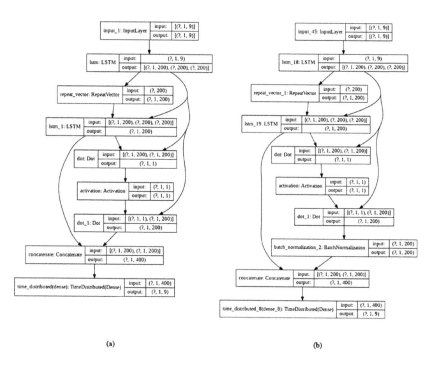

(a)　(b)

FIGURE 1.15 Model plot for the LSTM sequence-to-sequence (seq2seq) model with Luong attention: (a) Without batch normalization; (b) With batch normalization.

1.7.3 LSTM SEQUENCE-TO-SEQUENCE WITH BAHDANAU ATTENTION

Implementation of the Bahdanau attention mechanism is quite tricky. Though TensorFlow has made an attention layer available (tfa.seq2seq.Bahdanau attention), it mainly focuses on the translation kind of problem, and very low customization is possible. Hence in this research, a different approach is followed. A third-party implementation of Bahdanau attention [43, 44] is used here. Ahmed programs a custom attention layer, which is programmed for compatibility with Keras V 2.2.0. For the newer

```
inputs shape: (None, 1, 200)
Model: "functional_27"

Layer (type)              Output Shape         Param #
================================================================
input_49 (InputLayer)     [(None, 1, 9)]       0

lstm_20 (LSTM)            (None, 1, 200)       168000

attention_decoder_14 (Attent (None, 1, 9)      370090

time_distributed_9 (TimeDist (None, 1, 9)      90
================================================================
Total params: 538,180
Trainable params: 538,180
Non-trainable params: 0
```

```
Model: "functional_13"

Layer (type)              Output Shape                     Param #
==================================================================
encoder_inputs (InputLayer) [(None, 1, 9)]                 0

lstm_6 (LSTM)             [(None, 1, 200), (None, 2        168000

batch_normalization_5 (Batch (None, 200)                   800

repeat_vector_2 (Repeatvecto (None, 1, 200)                0

attention_decoder_2 (Attenti (None, 1, 9)                  370090

batch_normalization_7 (Batch (None, 1, 9)                  36

time_distributed_2 (TimeDist (None, 1, 9)                  90
==================================================================
Total params: 539,016
Trainable params: 538,598
Non-trainable params: 418
```

(a)   (b)

FIGURE 1.16 Model summary for LSTM sequence-to-sequence (seq2seq) with Bahdanau attention: (a) Without batch normalization; (b) With batch normalization.

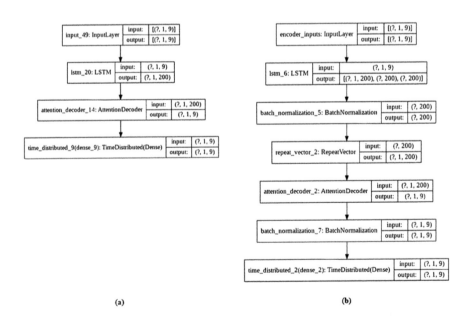

(a)   (b)

FIGURE 1.17 Model plots for LSTM sequence-to-sequence (seq2seq) with Bahdanau attention: (a) Without normalization; (b) With normalization.

versions of Keras, there is no separate Recurrent Layer Class. Hence, a separate base class code for the Recurrent Layer must be added from the old Keras repository and the modification in the attention layer is done. Batch normalization is applied to avoid gradient explosion caused by ELU or RELU [45] in the encoder. The original implementation was done without batch normalization; in this research, the performance of Bahdanau attention with and without batch normalization is studied. Accordingly, two different models are tested. Figure 1.16 shows the model summary for the implemented Bahdanau models, with and without batch normalization [46, 47], Figure 1.17 shows the plot for the implemented models; additional layers can be seen in Figure 1.17(b).

The following section is dedicated to the results; all four different models are trained on the 165 days of data and tested on 41 days of COVID-19 Data.

1.8 RESULTS AND DISCUSSION

For all the models mentioned above, implementation with and without batch normalization was performed. COVID-19 data of 208 was used out of that data of 166 days was used for training Purpose. For the testing phase, the remaining 41 days of data were used. Total 200 epochs with 5 as the batch size were done while the training phase was performed. The predictions of the seq2seq networks are overlaid on the actual plot. The results are shown below.

  (i) Regular LSTM (seq2seq)
 (ii) LSTM with Luong attention
(iii) LSTM with Bahdanau attention (custom attention layer)
(iv) LSTM with Bahdanau Attention-Keras Implementation

Figures 1.18 to 1.21 show the prediction of COVID-19 cases vs. the actual cases; the prediction is shown in red colour lines, it is observed that the LSTM models with attention mechanisms are having better approximation as shown in Figures 1.19 and 1.20. The Loss and Mean Absolute Error (MAE) [48] is given in Table 1.1.

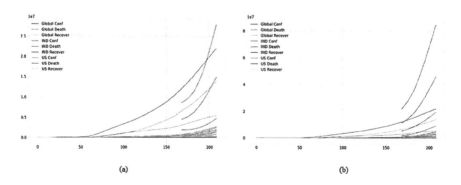

(a)   (b)

FIGURE 1.18 COVID-19 impact prediction results by regular LSTM without attention: (a) Without normalization; (b) With normalization.

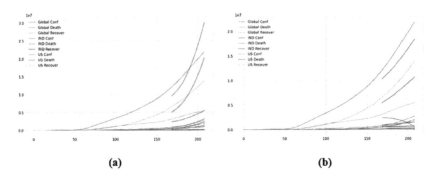

FIGURE 1.19 COVID-19 impact prediction results by LSTM with Luong Attention: (a) Without normalization; (b) With normalization.

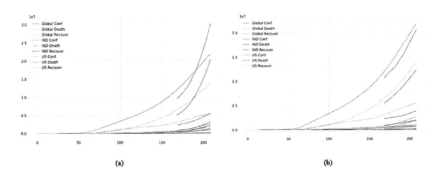

FIGURE 1.20 COVID-19 impact prediction results by LSTM with Bahdanau attention (custom attention layer): (a) Without normalization; (b) With normalization.

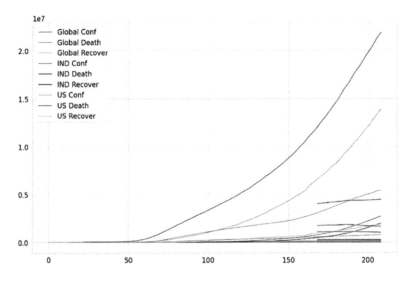

FIGURE 1.21 COVID-19 impact prediction results by LSTM with Bahdanau attention (KERAS attention layer).

TABLE 1.1

Performance Comparison of the Seq2Seq LSTM Models with Various Attention Mechanisms

Sr.	Model	Model Indicator	Loss	MAE
1	Regular LSTM with normalization	LSTM-R-WN	0.06	0.20
2	Regular LSTM without normalization	LSTM-R-N	3.20	1.34
3	LSTM with Luong Attention, with normalization	LSTM-A-L – N	0.07	0.19
4	**LSTM with Luong Attention, without normalization**	**LSTM-A-L – WN**	**0.03**	**0.14**
5	**LSTM with Bahdanau Attention, with normalization (custom layer)**	**LSTM-A-B-N**	**0.04**	**0.14**
6	LSTM with Bahdanau Attention, without normalization (custom layer)	LSTM-A-B-WN	0.18	0.39
7	LSTM with Bahdanau Attention, Keras attention layer	LSTM-A-B-KERAS	0.33	0.54

Note: Bold highlight best performance

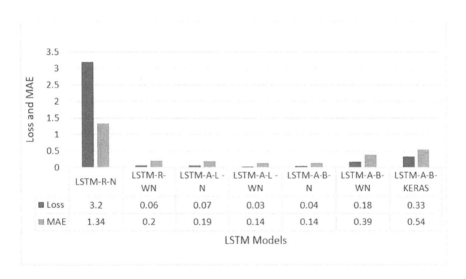

FIGURE 1.22 COVID-19 impact prediction results—Performance comparison of the seq2seq LSTM models with various attention mechanisms.

The Luong attention without batch normalization has given the best performance. Next best is given by LSTM with Bahdanau Attention, with batch normalization (custom layer).

Figure 1.22 shows the comparison chart; loss and mean absolute error are plotted for the various models implemented. It can be seen that LSTM models with attenuation have given the best performance.

The Python code and the dataset used for the research mentioned above are available for download and further extension on the link: http://doi.org/10.5281/zenodo.4288136 [49].

1.9   CONCLUSION

In this chapter, research work based on seq2seq LSTM networks is presented. The LSTMs with encoder-decoder architecture and attention mechanism is implemented. The attention mechanism enables LSTMs to understand the better past context for the prediction of the next values. The Luong and Bahdanau attention mechanism are described in this work. Both have their advantages and disadvantages, and their use will depend on the data to be used for the training purpose and the problem under consideration. The effect of batch normalization is studied, and it was found to be effective against the gradient explosion while a specific activation function such as 'ELU' is used. The attention mechanism results in lower loss and MAE while predicting the COVID-19 impact statistics.

REFERENCES

1. H. Ismail Fawaz, G. Forestier, J. Weber, L. Idoumghar, and P. A. Muller, "Deep learning for time series classification: A review," *Data Min. Knowl. Discov.*, vol. 33, no. 4, pp. 917–963, July 2019, doi: 10.1007/s10618-019-00619-1.
2. W. Aswolinskiy, R. F. Reinhart, and J. Steil, *"Time series classification in reservoir-and model-space: A comparison,"* in *Lecture Notes in Computer Science (Including Subseries Lecture Notes in Artificial Intelligence and Lecture Notes in Bioinformatics)*, vol. 9896 LNAI, pp. 197–208, 2016, doi: 10.1007/978-3-319-46182-3_17.
3. A. Bagnall, J. Lines, A. Bostrom, J. Large, and E. Keogh, "The great time series classification bake off: A review and experimental evaluation of recent algorithmic advances," *Data Min. Knowl. Discov.*, vol. 31, no. 3, pp. 606–660, May 2017, doi: 10.1007/s10618-016-0483-9.
4. H. Bouhamed and Y. Ruichek, "Deep feed-forward neural network learning using local binary patterns histograms for outdoor object categorization," *Adv. Model. Anal. B*, vol. 61, no. 3, pp. 158–162, Sep. 2018, doi: 10.18280/ama_b.610309.
5. D. C. Cires¸an, U. Meier, L. M. Gambardella, and J. Schmidhuber, "Deep, big, simple neural nets for handwritten digit recognition," *Neural Comput.*, vol. 22, no. 12. pp. 3207–3220, Dec. 1, 2010, doi: 10.1162/NECO_a_00052.
6. G. Wang, Q.-S. Jia, J. Qiao, J. Bi, and C. Liu, "A sparse deep belief network with efficient fuzzy learning framework," *Neural Networks.*, vol. 121, pp. 430–440, Jan. 2020, doi: 10.1016/j.neunet.2019.09.035.
7. S. Fort, H. Hu, and B. Lakshminarayanan, "Deep ensembles: A loss landscape perspective," Dec. 2019, Accessed: Aug. 25, 2020. [Online]. Available: http://arxiv.org/abs/1912.02757.
8. G. Liu and J. Guo, "Bidirectional LSTM with attention mechanism and convolutional layer for text classification," *Neurocomputing*, vol. 337, pp. 325–338, Apr. 2019, doi: 10.1016/j.neucom.2019.01.078.
9. A. Sherstinsky, "Fundamentals of recurrent neural network (RNN) and long short-term memory (LSTM) network," *Phys. D Nonlinear Phenom.*, vol. 404, Mar. 2020, doi: 10.1016/j.physd.2019.132306.

10. Z. C. Lipton, J. Berkowitz, and C. Elkan, "A critical review of recurrent neural networks for sequence learning," May 2015, Accessed: Nov. 5, 2020. [Online]. Available: http://arxiv.org/abs/1506.00019.

11. S. Hochreiter and Jürgen Schmidhuber, "*LSTM can solve hard long time lag problems*," in *NIPS'96: Proceedings of the 9th International Conference on Neural Information Processing Systems*, pp. 473–479, 1996, doi: 10.5555/2998981.2999048.

12. Y. Yu, X. Si, C. Hu, and J. Zhang, "A review of recurrent neural networks: LSTM cells and network architectures," *Neural Comput.*, vol. 31, no. 7, 1235–1270, Jul. 1, 2019, doi: 10.1162/neco_a_01199.

13. R. DiPietro and G. D. Hager, "Deep learning: RNNs and LSTM," in *Handbook of Medical Image Computing and Computer Assisted Intervention*, Elsevier, 2019, pp. 503–519.

14. J. H. Bappy, C. Simons, L. Nataraj, B. S. Manjunath, and A. K. Roy-Chowdhury, "Hybrid LSTM and encoder-decoder architecture for detection of image forgeries," *IEEE Trans. Image Process.*, vol. 28, no. 7, 3286–3300, Jul. 2019, doi: 10.1109/TIP.2019.2895466.

15. S Hochreiter and J. Schmidhuber "Long short-term memory," *Neural Comput.*, vol. 9, no. 8, 1735–1780, 1997, doi: https://doi.org/10.1162/neco.1997.9.8.1735.

16. V. Veeriah, N. Zhuang, and G. J. Qi, "*Differential recurrent neural networks for action recognition*," in *2015 IEEE International Conference on Computer Vision (ICCV)*, pp. 4041–4049, Santiago, Chile, 2015, doi: 10.1109/ICCV.2015.460.

17. P. Zhao, H. Zhu, Y. Liu, Z. Li, J. Xu, and V. S. Sheng, "Where to go next: A spatio-temporal LSTM model for next POI recommendation," *IEEE Trans. Knowl. Data Eng.*, July 2020, doi: 10.1109/TKDE.2020.3007194

18. R. Jozefowicz, W. Zaremba, and I. Sutskever, "An empirical exploration of recurrent network architectures," in *ICML'15: Proceedings of the 32nd International Conference on International Conference on Machine Learning.* vol. 37, pp. 2342–2350, July 2015.

19. H. Xu, C. Zhang, G. S. Hong, J. Zhou, J. Hong and K. S. Woon, "*Gated recurrent units based neural network for tool condition monitoring*," in *2018 International Joint Conference on Neural Networks (IJCNN)*, pp. 1–7, 2018, doi: 10.1109/IJCNN.2018.8489354.

20. J. Kim, M. El-Khamy, and J. Lee, "Residual LSTM: Design of a deep recurrent architecture for distant speech recognition," 2017, Arxiv, doi: 10.21437/Interspeech.2017-477.

21. R. Pascanu, T. Mikolov, and Y. Bengio, "*On the difficulty of training recurrent neural networks*," in *International Conference on Machine Learning*, vol. 28, 17–19 June 2013, Atlanta, GA, 2013.

22. S. Bengio, O. Vinyals, N. Jaitly, and N. Shazeer, "*Scheduled sampling for sequence prediction with recurrent neural networks*," in *NIPS'15: Proceedings of the 28th International Conference on Neural Information Processing Systems*, vol. 1, pp. 1171–1179, December 2015, Montreal, Canada.

23. A. M. Fernandez-Escamilla, F. Rousseau, J. Schymkowitz et al. "Prediction of sequence-dependent and mutational effects on the aggregation of peptides and proteins," *Nat. Biotechnol.*, vol. 22, pp. 1302–1306, 2004. https://doi.org/10.1038/nbt1012

24. M. Wang, Z. Lu, H. Li, W. Jiang, and Q. Liu, "A convolutional architecture forward sequence prediction," arXiv:1503.03244 [cs.CL] 2015.

25. Z. Xing, J. Pei, and E. Keogh, "A brief survey on sequence classification," *ACM SIGKDD Explor. Newsl.* vol. 12 no. 1, pp 40–48, June 2010. https://doi.org/10.1145/1882471.1882478

26. C. C. Aggarwal, *"On effective classification of strings with wavelets,"* in *KDD '02: Proceedings of the Eighth ACM SIGKDD International Conference on Knowledge Discovery and Data Mining*, Edmonton, Alberta, Canada, pp. 163–172, July 2002, doi: 10.1145/775047.775071.

27. J. Lin, E. Keogh, L. Wei et al. Experiencing SAX: A novel symbolic representation of time series. *Data Min. Knowl. Disc.* vol. 15, pp. 107–144, 2007. https://doi.org/10.1007/s10618-007-0064-z

28. D. Heller, R. Krestel, U. Ohler, M. Vingron, and A. Marsico, "ssHMM: extracting intuitive sequence-structure motifs from high-throughput RNA-binding protein data," *Nucleic Acids Res.*, vol. 45, no. 19.2, pp. 11004–11018, November 2017. https://doi.org/10.1093/nar/gkx756

29. S. Wiseman and A. M. Rush, *"Sequence-to-sequence learning as beam-search optimization,"* in *Proceedings of the 2016 Conference on Empirical Methods in Natural Language Processing*, Austin, TX, pp. 1296–1306, 2016, doi: 10.18653/v1/d16-1137.

30. G. Kurata, B. Xiang, and B. Zhou, *"Labeled data generation with encoder-decoder LSTM for semantic slot filling,"* in *INTERSPEECH 2016*, pp. 8–12 Sep. 2016, San Francisco, CA, doi: 10.21437/Interspeech.2016-727.

31. Y. Zhang, D. Li, Y. Wang, Y. Fang, and W. Xiao, "Abstract text summarization with a convolutional seq2seq model," *Appl. Sci.*, vol. 9, no. 8, 1665. 2019, doi: 10.3390/app9081665.

32. G. Liu and J. Guo, "Bidirectional LSTM with attention mechanism and convolutional layer for text classification," *Neurocomputing*, vol. 337, no. 14 pp. 325–333, April 2019, doi: 10.1016/j.neucom.2019.01.078.

33. J. H. Bappy, C. Simons, L. Nataraj, B. S. Manjunath, and A. K. Roy-Chowdhury, "Hybrid LSTM and encoder-decoder architecture for detection of image forgeries," *IEEE Trans. Image Process.*, vol. 28, no. 7, pp 3286–3300 2019, doi: 10.1109/TIP.2019.2895466.

34. L. Wu et al., *"Adversarial neural machine translation,"* in *Proceedings of Machine Learning Research: The 10th Asian Conference on Machine Learning*, Beijing, vol. 95, pp. 534–549, 2018.

35. D. Bahdanau, K. H. Cho, and Y. Bengio, *"Neural machine translation by jointly learning to align and translate,"* in *3rd International Conference on Learning Representations, ICLR 2015*, San Diego, CA, pp 1–15, arXiv:1409.0473 [cs.CL], 2015.

36. M. T. Luong, H. Pham, and C. D. Manning, "Effective approaches to attention-based neural machine translation," *AI Open* vol. 1, pp. 5–21, 2020, doi: 10.18653/v1/d15-1166.

37. H. Li, Z. Liu, and J. Ge, "Scientific research progress of COVID-19/SARS-CoV-2 in the first five months," *J. Cell. Mol. Med.* vol. 24, no. 12, pp. 6558–6570, June 2020, doi: 10.1111/jcmm.15364.

38. E. Dong, H. Du, and L. Gardner, "An interactive web-based dashboard to track COVID-19 in real time," *Lancet Infect. Dis.* vol. 20, no. 5, pp. 533–534, 2020, doi: 10.1016/S1473-3099(20)30120-1.

39. M. Bogner, F. Weindl, and F. Wiesinger, *"Software frameworks for artificial intelligence: Comparison of low-level and high-level approaches,"* in *Proceedings of the 31st European Modeling and Simulation Symposium EMSS2019*, Lisbon, Portugal, pp. 96–102, 2019, ISBN: 9788885741263, doi: 10.46354/i3m.2019.emss.016.

40. F. Chollet, Keras, GitHub, 2015. Available at: https://github.com/fchollet/keras.

41. M. Abadi et al., *"TensorFlow: A system for large-scale machine learning,"* in *12th symposium on operating systems design and implementation*. Savannah, GA, pp. 265–283, November 2–4, 2016.

42. X. Shi, Z. Chen, H. Wang, D. Y. Yeung, W. K. Wong, and W. C. Woo, "*Convolutional LSTM network: A machine learning approach for precipitation nowcasting*," in *Advances in Neural Information Processing Systems 28 (NIPS 2015)*, arXiv:1506.04214 [cs.CV], 2015.
43. Z. Ahmed, "How to visualize your recurrent neural network with attention in Keras," *Medium*, Online Article - https://medium.com/datalogue/attention-in-keras-1892773a4f22, 2017.
44. F. Karim, S. Majumdar, H. Darabi, and S. Chen, "LSTM fully convolutional networks for time series classification," *IEEE Access*, vol. 6, pp. 1662–1669, 2017, doi: 10.1109/ACCESS.2017.2779939.
45. A. F. M. Agarap, "Deep learning using rectified linear units (ReLU)," arXiv. 2018.
46. S. Singh and S. Krishnan, "Filter response normalization layer: Eliminating batch dependence in the training of deep neural networks," arXiv:1911.09737v2 [cs.LG], 2020,.
47. X. Li, S. Chen, X. Hu, and J. Yang, "*Understanding the disharmony between dropout and batch normalization by variance shift*," in *2019 IEEE/CVF Conference on Computer Vision and Pattern Recognition (CVPR)*, Long Beach, CA, pp. 2677–2685, June 15–20, 2019, doi: 10.1109/CVPR.2019.00279.
48. T. Chai and R. R. Draxler, "Root mean square error (RMSE) or mean absolute error (MAE)?-Arguments against avoiding RMSE in the literature," *Geosci. Model Dev.*, 2014, doi: 10.5194/gmd-7-1247-2014.
49. V. A. Bharadi, S. S. Alegavi, "Using Luong and Bahdanau attention mechanism on the long short term memory networks—A COVID-19 impact prediction case study (Online Code and Data Repository - version 1.0)," *Zenodo*, 2020, http://doi.org/10.5281/zenodo.4288136

2 A Novel Missing Data Imputation Algorithm for Deep Learning-Based Anomaly Detection System in IIoT Networks

Ancy Jose and S.V. Annlin Jeba
Sree Buddha College of Engineering, Kerala, India

Beulah Joslyn Jose
Temple University

CONTENTS

2.1 INTRODUCTION

The recent advancements in technology pave the way to bring drastic changes in all the fields. 'Smart Computing' is an in-demand term that plays a vital role in this modern era. Smart computing is an emerging technology that describes the integration of hardware, software and network techniques. This technology provides the IT frameworks with real-time understanding and comprehensive investigations of the real world. This technology allows individuals to make more smart conclusions about alternatives and actions that strengthen business operations. This technology is in a higher interest these days and enlarges at a higher pace in upcoming years. 'Self-Adaptive System' is another term playing a significant role in the current technological era. Self-Adaptive systems are the systems that can automatically modify themselves with respect to changes in their system behavior. They can modify their runtime behavior to achieve system goals.

The Industrial Internet of Things (IIoT) is a popular smart computing and self-adaptive technology. IIoT is a major technology in the current digitization revolution. IIoT is a subcategory of the Internet of Things (IoT). IoT is a collection of interconnected devices, whereas, IIoT refers to IoT technology in industrial settings and business operations. IIoT technologies are transforming the way manufacturing organizations are run. It enables industrial operations to be more productive and efficient. Due to this enhanced productivity and higher efficiency, industrial settings increasingly rely on IIoT devices and applications. Due to this increased reliability on IIoT devices and applications, they are always vulnerable to cyber-attacks and major threats. This leads to severe loss and major problems in industrial processes as well as business operations. Some of the cyber threats that IIoT infrastructure is exposed to include man in the middle, device hijacking, Distributed Denial of Service (DDoS), etc. IIoT substructure must be secured by a lasting security solution that does not interrupt its functional operations, service reliability, or advantages.

To protect CI like IIoT, Anomaly Detection Systems (ADSs) are usually implemented. ADSs are developed using some statistical methods that identify what patterns are normal and then find items that do not imply those patterns. The major purpose of the ADS systems is to efficiently and effectively classify the normal and malicious packets or data records from the input data transmitting through the network. The most popular methods to develop ADS systems are machine learning and deep learning-based techniques. Several ADSs had been proposed and implemented in the existing systems. But the mechanisms proposed in existing literatures detect only the intrusions and anomalies occurring in the IIoT sector. These do not implement any technique to validate the effectiveness of the input datasets used in these ADS systems. The input dataset used in these detection systems must be proper and valid to build accurate detection models.

A dataset with missing values or missing features cannot be processed to build an effective ADS. It can cause adverse effects in the detection results and reduce the model's strength and accuracy. If an ADS system is processed with an input dataset with missing values, it leads to the wrong classification of data and biased results. It can have drastic effects on the output that is obtained from the input data given. The ADS system may classify normal data as malicious and malicious data as normal. The data labeled as normal by the ADS system will be used for further processing in the IIoT network. Ultimately, this wrong classification of data leads to a major problem like cyber threats. Hence, missing values in the dataset is one of the serious problems to be addressed. Therefore, it is important to fill in the missing values in the dataset before the ADS system is processing it.

Current ADS systems, especially deep learning-based ADS systems, do not efficiently handle the missing values in input data. This missing value problem could be handled only by the missing data imputation process. Missing data imputation must be done in the phase of data preprocessing to reduce the possibility of missing data. So, effective missing data imputation techniques must be introduced to protect CI like IIoT.

The contributions of this chapter are as follows. A new type of deep learning-based ADS system is proposed in this chapter. The proposed ADS system implements a novel missing data imputation algorithm named Enhanced DNN for Missing Data Imputation (EDMDI) to handle the problems caused by missing values in input data. The proposed algorithm imputes the possible values in the missing fields in the input dataset. The newly created imputed dataset will be used for further processing in the ADS system. The proposed ADS system is implemented using deep learning techniques. This work mainly focuses on developing an algorithm to automatically fill the missing values in the dataset before it is processed. This algorithm imputes the missing values in the dataset-based on the existing data evidences. The proposed EDMDI-based ADS system can accurately identify the malicious activities occurring in the IIoT network and help to reduce the network's vulnerability to major cyber threats.

The remaining sections of this chapter are organized as follows. Section 2.2 provides an overview of the existing ADS systems. Section 2.3 demonstrates the proposed ADS system. Section 2.4 illustrates a theoretical analysis performed on existing and proposed systems. Section 2.5 discusses the experimental results obtained when the proposed ADS system is processed with a well-known network dataset. Section 2.6 concludes the chapter with future scope.

2.2   AN OVERVIEW ON EXISTING ADS SYSTEMS

Different techniques have been proposed to identify anomalies in internet traffic. Researchers have proposed various machine learning and deep learning techniques to identify malicious activities occurring in network traffic. This section describes a review conducted on existing ADS systems and the imputation methods proposed by different researchers to impute the missing data in the input dataset used in these ADS systems.

Astuti et al. [1] analyzed the performances of enforcing the different number of cross-fold validations of missing data imputation techniques. The results obtained do not exhibit notable improvements. Robert Mitchell et al. [2] proposed a behavior-rule specification-related method for anomaly detection of medical appliances implanted in a Medical Cyber-Physical System (MCPS). Vital sign monitor medical appliances are taken as an example to prove that the proposed intrusion detection method can successfully switch false positives off for a high detection probability to handle hidden attackers. Comparative analysis proves that this technique performs better than two existing anomaly-based techniques for identifying unusual patient behaviors in ubiquitous healthcare applications.

Zhun-Ga Liu et al. [3] proposed a novel prototype-dependent Credal Classification (PCC) technique to compete with incomplete patterns based on evidential reasoning method. A new credal combination technique is presented for rectifying the classification problem. It is possible to distinguish the intrinsic uncertainty due to the feasible contradicting results provided by various findings of the missing values. The efficacy of this method is tested using four experiments with artificial and real-time datasets.

Mohammed A. Ambusaid et al. [4] proposed a mutual information-based algorithm to select the best feature for classification. An Intrusion Detection System (IDS) called Least Square Support Vector Machine-based IDS (LSSVM-IDS) is developed based on the features extracted by the proposed feature selection algorithm. The evaluation results demonstrate that the proposed feature selection algorithm achieves better accuracy. Hamed Haddad Pajouh et al. [5] developed a novel model for intrusion detection depending on two-layer dimension reduction and two-tier classification module. The proposed model uses component and linear discriminant analysis to convert the high-dimensional dataset to a lower one with fewer features. The experiment uses the NSL-KDD dataset and results depict that the model performs better than previous models.

Zach DeSmit et al. [6] proposed an approach for systemically finding cyber threats. The intersection mapping concept is used to find vulnerabilities in the manufacturing sector. A case study of the newly proposed approach is done at an applied manufacturing research facility. Muna AL-Hawawreh et al. [7] proposed an anomaly detection method based on a deep learning model to identify anomalies occurring in industrial IoT. Deep autoencoder and deep feed-forward neural network are used. Evaluation is done in the NSL-KDD dataset. Results prove that this system has a higher detection rate.

Xavier Clotet et al. [8] proposed a real-time anomaly-based detection system to work at the industrial operation level of CI. The proposed IDS uses a multi-agent method to handle the difficult problem of examining huge amounts of data obtained from dimensions registered by ICS. The proposed tool was developed, validated and verified throughout the EU-funded project PREEMPTIVE. Detection results are observed on the data collected from a water treatment plant laboratory.

Jose Francisco Colom et al. [9] proposed a scheduling framework for distributed ADSs over heterogeneous architectures. The proposed framework contains two components namely a controller component and a switching mechanism. This approach is tested through different experiments. The experiments conducted showed the feasibility of this approach.

Maede Zolanvari et al. [10] describe an experiment conducted to evaluate why machine learning must be fused into the IIoT security mechanisms. The problems and real-world analysis embedded with this case are studied in the experimental outline. An IIoT testbed indicating a real industrial plant is also used to prove the concept. Amjad Mehmood et al. [11] proposed a new Naive Bayes classification algorithm in IDSs. This work mainly concentrates on securing an IoT infrastructure from distributed denial-of-service attacks performed by attackers.

Muhammad Wasimuddin et al. [12] proposed a Branching Artificial Neural Ensemble (BRANE) to rectify missing values in attributes. It is an addendum of the

Ensemble method. It uses one multilayer perceptron for data points, one multilayer perceptron for missing values as a flag entity and one multilayer perceptron to fuse the experts and forecast the final result. Zillow Competition dataset from the Kaggle database is used in this method.

Berihun Fekade et al. [13] proposed a probabilistic method to retrieve missing data from IoT sensors. Simulation results prove that the proposed PMF model with clustering performs better than Support Vector Machine (SVM) and Deep Neural Network (DNN) algorithms concerning accuracy and root mean square error. Cheng-Bo Lu et al. [14] developed an imputation method for missing data using an Extreme Learning Machine Auto-Encoder (ELM-AE). The performance of ELM-AE imputation is compared with some other popularly used imputation methods using seven UCI benchmark datasets. The proposed ELM-AE imputation method outperforms the other three techniques.

Xiang Yu et al. [15] developed a new multi-attribute missing data reconstruction method to rebuild the lost data with increased accuracy. This technique is evaluated in a real sensor dataset obtained from the Intel Berkeley Research Laboratory, consisting of two missing patterns: random missing pattern and continuous missing pattern. Result obtained is that this performs well even to fill large missing regions.

Tiago Mazzutti et al. [16] proposed an Incremental Neuro-Fuzzy Gaussian Mixture Network (INFGMN) method using an approximated enhanced version of the Expectation-Maximization (EM) algorithm for carrying out the imputation stage of the missing data throughout the implementation of recalling in the network layer of the INFGMN, making it possible to deal with missing data. The modeling performance of the INFGMN in the occurrence of missing data is validated using various benchmark applications. It is concluded that the proposed model is a feasible substitute to the existing ones for the data filling.

Alessio Petrozziello et al. [17] proposed a framework for data filling in Spark, permitting easy imputation as a subsidiary step to the pre-processing data stage. The developed one is tested on a real-world Recommender Systems dataset, where the missing data is commonly a problem for new items. The model's ranking is usually prejudiced in the direction of the popular items.

Shamsul Huda et al. [18] proposed a malicious threat-detection model using a deep belief network for Cloud-assisted Internet of Things (CoT)-based Industrial Control System (ICS). The proposed detection models are substantially validated on a real malware testbed. Experimental results demonstrate that the proposed method achieves higher accuracies than existing detection algorithms and portrays similar performances with previous semi-supervised works. Ivan Izonin et al. [19] proposed a regression-based methodology for solving the activity of incomplete data retrieval. Among several regression approaches, the developed technique fills in accurate missing data.

Daming Li et al. [20] developed an IoT feature extraction and anomaly detection method for the smart city using a deep migration learning model. The experimental evaluations show that the proposed system has a lesser detection time and higher detection rate. J. Kirupakar et al. [21] proposed a smart architectural prototype for the constrained IIoT portals to find cyber threats in the Industrial IoT area effectively. Maede Zolanvari et al. [22] introduced a machine learning-based network vulnerability analysis. At first common IIoT protocols and their related threat vulnerabilities are presented. Secondly, a cyber-vulnerability assessment is conducted and the usage of ML in countering these vulnerabilities is discussed. Finally, a case study including details of a real-world testbed is performed. The performance is also evaluated based on representative metrics to demonstrate the effectiveness of these methods.

Mahmudul Hasan et al. [23] analyzed various machine learning-based attack detection models for IoT. Machine learning algorithms used for comparison are Logistic Regression (LR), SVM, decision tree, random forest and Artificial Neural Network (ANN). Evaluation results show that 99.4% accuracy has been achieved by a decision tree, random forest and ANN. Chen et al. [24] proposed Fiden to study the attacker identification problem in the IIoT. A real case study demonstrates the applicability of Fiden and the results prove that the proposed algorithm identifies the devices that launched attacks. Yao et al. [25] proposed a hybrid IDS architecture and a machine learning-based detection mechanism. This performs greater in comparison with the existing literatures with respect to a scale of benchmarks.

Several kinds of literature on ADS systems are analyzed and the following findings are obtained. Different types of IDSs are proposed in existing literature. The methods specified in each literature detect intrusion occurring in the IIoT network and protect this network against various kinds of attacks. Most of the methods depend on machine learning algorithms like SVM, decision tree, random forest, etc. But existing literature do not discuss the problem of handling missing data in IDSs, especially in deep learning-based ones. Missing values in the dataset can cause degradation of the system performance. So proper techniques and algorithms must be introduced to rectify the missing value problems in the input dataset used.

2.3   PROPOSED ADS SYSTEM

2.3.1   Overview of the Proposed EDMDI Algorithm-Based ADS System

ADSs are implemented in CI networks to protect confidential data from cyber-attacks. The dataset passing through the network is given as input to the ADS systems and these systems classify the normal and malicious data records. But the ADS systems produce biased outcomes if the input dataset given to them is processed with missing values and features. The missing data in the input dataset leads to the wrong classification of normal and abnormal data records by the ADS system. To solve this missing data problem, a new type of ADS system is proposed. The proposed ADS system initially checks for any missing fields in the input dataset and then imputes the missing fields with possible values. A new algorithm named Enhanced DNN for Missing Data Imputation (EDMDI) is proposed to perform missing data imputation. After the imputation process, the imputed dataset is used to process the ADS system further to identify if any malicious packets or data are present. The proposed ADS system performs malicious data detection using deep learning techniques. If the ADS system detects any malicious packets in the imputed input dataset, a warning will be provided by the ADS system that malicious data is detected. In this way, an ADS system is built and deployed in the IIoT network to identify the malicious activities occurring in the network.

2.3.2 Architecture of the Proposed Scheme

The below mentioned are the important stages in the proposed ADS system. The proposed ADS system is implemented in the following steps: (a) Missing Data Imputation using EDMDI algorithm, (b) ADS Model Creation using Supervised Learning, and (c) Malicious Data Detection using DNN.

a. **Missing Data Imputation Using EDMDI Algorithm**
   This ADS model implements an algorithm called EDMDI. This algorithm will check for any missing fields in the input dataset given to the ADS system. If any missing fields are present, EDMDI will impute the missing fields with possible values and finally an imputed dataset will be created. This new dataset will be used to process this ADS system further to classify the normal and malicious packets. Since the DNN classifier is used for classification by this ADS system, a new deep learning layer is created in this ADS system to process the EDMDI algorithm. The ADS model has been trained using several deep learning layers. Of these layers, the initial layer is the newly created layer. This layer is created to process the newly developed EDMDI algorithm.

b. **ADS Model Creation Using Supervised Learning**
   Initially the input dataset taken is split into a training dataset and testing dataset. The training dataset is used to train the model. This is done to create the ADS model. This stage involves the training stage. All the data used here are labeled data. So supervised learning is carried out to train the model. Supervised learning is carried out using the deep learning technique. DNN is used to train the model. In this stage, finally, a trained anomaly detection model will be obtained. This model will be trained in such a way that this can efficiently classify the normal and anomaly data.

c. **Malicious Data Detection Using DNN**
   In this stage, the test set is loaded to the developed ADS model. This dataset may contain missing values. So in this stage, the dataset is processed initially in the newly developed layer. The missing values are imputed by the newly developed algorithm in this layer. This layer produces a new imputed dataset. This new dataset will be used for further processing/classification in the ADS system. If any malicious packets are found, the ADS model provides an alert that malicious packets are detected. Then the malicious traffic can be discarded from the network.

Figure 2.1 illustrates the architecture of the proposed ADS system. Initially, the training set will be loaded and based on supervised learning using DNN, the model will be trained and the ADS model will be created. After model creation, the test set is loaded and checked for missing fields. If missing fields are present, the newly developed EDMDI algorithm imputes the missing fields with possible values and creates a new imputed test set. Then the imputed test set will be used as an input dataset to the ADS system. Then the classifier detects malicious data from the input set using the DNN technique and classifies the data as normal and anomaly accordingly. Here the EDMDI algorithm is processed in a newly created deep learning layer.

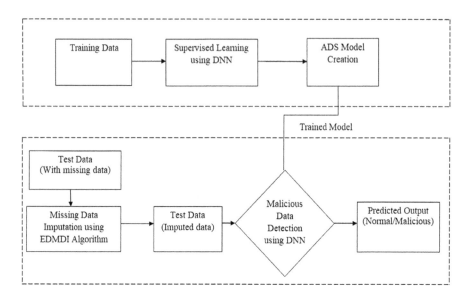

FIGURE 2.1   Architecture of proposed ADS system.

2.3.3   ENHANCED DNN FOR MISSING DATA IMPUTATION ALGORITHM

The proposed EDMDI algorithm computes the possible values that can be imputed in the missing fields based on the existing values from the input dataset. The possible values imputed in the missed fields are computed based on exact attribute combinations and like attribute combinations. Table 2.1 is a limited version of the input dataset used to conduct experiments to validate the performance and efficiency of the proposed algorithm. For instance, Table 2.1 has '4' data records, and if suppose the value of the attribute 'duration' in the first data record is missing, then the algorithm initially checks for exact attribute combinations, i.e., all the possible missed attribute values. Here, since there are '4' data records, for the missing attribute 'duration', the possible missed attribute values that can be imputed in the missing field are '4', i.e., '0', '2', '0' and '0'. The algorithm imputes '0' in missing fields initially. All these values will be stored in a table by the algorithm. In the final stage, the highly probable value will be imputed.

TABLE 2.1
Sample Dataset

Sl. No	Duration	Protocol_type	Service	Flag
1	0	Tcp	Private	REJ
2	2	Tcp	Private	REJ
3	0	Tcp	ftp	SF
4	0	Tcp	ftp	RSTO

Next, the algorithm looks for attribute combinations. Here, if suppose in the third data record, the missed attribute is 'service', then, the algorithm looks for the like attribute combinations. The third and fourth data records have the same attribute values for 'duration' and 'protocol_type'. These are called as like attribute combinations. The algorithm considers 'ftp' to be one of the possible values in the missing fields and stores in the table. Finally, from the identified possible missed attribute values of both the exact and like combinations, the highly probable value will be imputed in the missing field by the algorithm.

The key procedure of the EDMDI algorithm is as follows.

ALGORITHM 1   ENHANCED DNN FOR MISSING DATA IMPUTATION (EDMDI)

1: **Input:** Dataset with missing values D
2: **Output:** Imputed dataset I
3: Begin
4:    READ D
5:    CREATE datatb_m
6:    Mark tuple_id and column_id
7:    Find all the possible non-missing attribute_comb
8: Find all the possible missed attribute_value
9: SELECT * from datatb_m
10:    Calculate count of each missed attribute_value
11:        COMPUTE $prb = pc/tc$
12:    Count the number of attribute-value combinations
13:    Find attribute of marked tuple_id and column_id
14: SELECT * from attri_value_tb
15:    COMPUTE probable value
16: SELECT * from sel_comb
17:    COMPUTE the possible value based on sel_comb
18:    WRITE computed value in marked tuple_id and column_id
19:    Create I
20: End

Figure 2.2 illustrates the process of missing data imputation. Dataset with missing values is used as the input dataset. A new deep learning layer is developed to process the EDMDI algorithm. The algorithm scans the dataset and processes five stages to impute the possible values in the missing fields. All five stages are processed and the values are stored in a table in each stage. Initially, in stage 1, the algorithm finds the corresponding tuple id and column id of the missing fields and marks them in the table. In stage 2, the algorithm processes four modules to impute the missing fields. In module 1, all the possible missed attribute values are identified from the marked column_id. In module 2, the count of each possible missed attribute value is made and the probability of each identified possible missed attribute value is computed.

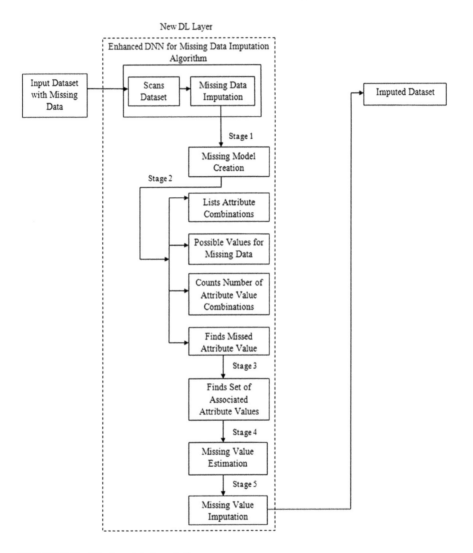

FIGURE 2.2 Missing data imputation process.

In module 3, the like attribute combinations are identified and the count of each like attribute combination is stored in the table. In module 4, the missed attribute is found. In stage 3, highly probable attribute combinations are selected from the identified exact and like attribute combinations. In stage 4, the possible missing value with high probability is computed from the selected attribute combinations. In stage 5, the value with the highest probability is imputed in the missing field. After the imputation process, a new imputed dataset is created. This new dataset is used for further processing in the ADS system. This imputed dataset will be used as an input dataset to the ADS system. Then the classifier detects malicious data from the input set using the DNN technique and classifies the data as normal and anomaly accordingly.

2.4 THEORETICAL ANALYSIS

Table 2.2 illustrates an analysis performed between the existing ADS system and the proposed EDMDI algorithm-based ADS system. The analysis is performed based on the factors such as accuracy, precision, recall, time complexity, attack detection rate, security, bias in estimates and efficiency. The existing ADS system implements only the DNN classifier to build an ADS system. It does not implement any technique or algorithm to handle the problems caused by missing data. But the proposed ADS system implements an EDMDI algorithm to handle the missing value problems in the dataset. The proposed system shows an overall improvement in performance with respect to accuracy. ADS system processed with missing data can cause a bias in output. But ADS system processed with an imputed dataset classifies the normal and abnormal data more accurately. So accuracy will be higher in the proposed system. Since classification is more accurate, the total number of instances retrieved will increase in the proposed system. So the precision and recall rate will be higher in the proposed ADS system. The EDMDI algorithm is developed in an effective manner such that the time taken to predict the output will be lesser when compared with the existing system. So time complexity is lesser in the proposed system. More number of attacks are detected by the proposed ADS system than the existing ADS system. So attack detection rate will also be higher in the proposed ADS. Since more attacks are identified, security is also higher in the proposed ADS system. Moreover, the proposed system is more efficient and also produces unbiased estimates than the existing system.

2.5 RESULTS AND DISCUSSION

2.5.1 DATASETS

The efficiency of the proposed algorithm is estimated on a well-known network dataset called NSL-KDD. NSL-KDD is one of the popular datasets used for building IDSs. After preprocessing, the NSL-KDD dataset contains a reasonable amount of 148,517 records with 41 features and class labels. Class labels indicate whether the data is normal or anomaly (unusual) data. These datasets are the records of internet traffic found by an intrusion detection network. This dataset consists of 77,054 normal and 71,460 attack data.

TABLE 2.2
Analysis on Existing ADS System and Proposed ADS System

Technique	Existing ADS System	Proposed ADS System
Accuracy	Lesser than imputed set	Higher than missing set
Precision	Comparatively less	Comparatively higher
Recall	Comparatively less	Comparatively higher
Time complexity	More processing time	Comparatively lesser processing time
Attack detection rate	Comparatively less	More attacks are detected
Security	Comparatively less	Higher
Bias in estimates	Biased estimates	Unbiased estimates
Efficiency	Comparatively less	More efficient than DNN-based ADS

2.5.2 EVALUATION METRICS

The proposed EDMDI algorithm-based ADS system is evaluated in terms of accuracy, precision, recall, time complexity and attack detection rate. These performance metrics are calculated as follows.

Accuracy: The accuracy of a system is the number of observations exactly found with reference to the total number of observations.

$$Accuracy = TP + TN / TP + TN + FP + FN$$

where TP = True Positive, TN = True Negative, FP = False Positive, FN = False Negative.

Time complexity: Time complexity is the time taken by an algorithm to run as a function of the length of the input. The lesser the time complexity, the more is the efficiency of the algorithm.

Attack detection rate: Attack detection rate is the total number of attacks identified.

Precision: Precision is the fraction of related instances among the obtained instances.

$$Precision = TP / TP + FP$$

Recall: Recall is the fraction of the total amount of related instances that are obtained.

$$Recall = TP / TP + FN$$

2.5.3 EXPERIMENTAL RESULTS

To validate the performance of the newly developed algorithm, missing fields are simulated in the input dataset and imputation is performed. The performance of the algorithm is validated using a different combination of records from the original dataset. The proposed algorithm is evaluated using the following combination of records. Different combinations of records like 100, 200, 500 and 1000 records are used as input datasets to train the ADS. Datasets with 10, 50 and 100 records are used for testing purposes. Full dataset usage leads to complex processing. Due to the processing limitations of the system, these combinations of dataset records are used. Missing fields are simulated in the test sets, and missing data imputation is performed in these datasets by the proposed algorithm.

The following graphs represent the results obtained after performing the anomaly data classification in the newly created imputed dataset. Figure 2.3 represents the time complexity analysis of DNN and I_DNN when the proposed ADS system is processed with 500 training data records against the different number of testing data records like 10, 50 and 100. DNN is a Deep Neural Network (with missing data) and I_DNN is Impute-Enabled DNN (imputed dataset). Figure 2.4 represents the time complexity analysis of DNN and I_DNN when the proposed ADS system is processed with 1000 training data records against the different number of testing data records like 10, 50 and 100.

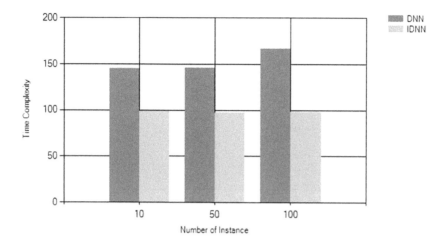

FIGURE 2.3   Time-complexity analysis of DNN vs I_DNN with 500 train records against 10, 50, 100 test records.

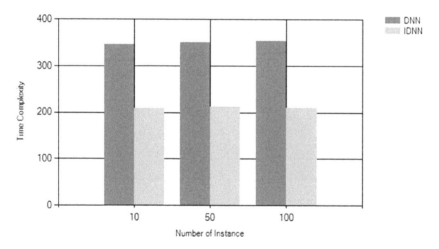

FIGURE 2.4   Time-complexity analysis of DNN vs I_DNN with 1000 train records against 10, 50, 100 test records.

Figure 2.5 represents the accuracy analysis of DNN and I_DNN when the proposed ADS system is processed with 500 training data records against the different number of testing data records like 10, 50 and 100. For 500 training records with missing data, the obtained accuracy is 96.5 % and with imputed data, the obtained accuracy is 96.6 %. Figure 2.6 represents the accuracy analysis of DNN and I_DNN when the proposed ADS system is processed with 1000 training data records against the different number of testing data records like 10, 50 and 100. For 1000 training records with missing data, the obtained accuracy is 89.1 % and with imputed data, the obtained accuracy is 97.1 %.

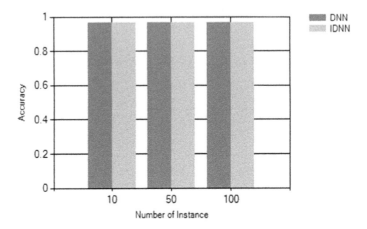

FIGURE 2.5   Performance of DNN vs I_DNN in terms of accuracy with 500 train records against 10, 50, 100 test records.

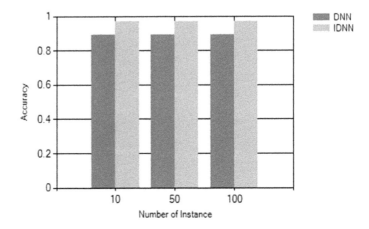

FIGURE 2.6   Performance of DNN vs I_DNN in terms of accuracy with 1000 train records against 10, 50, 100 test records.

Figure 2.7 represents the precision analysis of DNN and I_DNN when the proposed ADS system is processed with 500 training data records against the different number of testing data records like 10, 50 and 100. For 500 training records with missing data, the obtained precision is 95.6 % and with imputed data, the obtained precision is 96.3 %. Figure 2.8 represents the precision analysis of DNN and I_DNN when the proposed ADS system is processed with 1000 training data records against a different number of testing data records like 10, 50 and 100. For 1000 training records with missing data, the obtained precision is 88.3 % and with imputed data, the obtained precision is 97.1 %.

Figure 2.9 represents the recall rate analysis of DNN and I_DNN when the proposed ADS system is processed with 500 training data records against a different

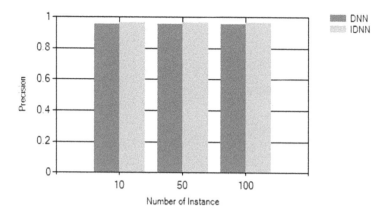

FIGURE 2.7 Performance of DNN vs I_DNN in terms of precision with 500 train records against 10, 50, 100 test records.

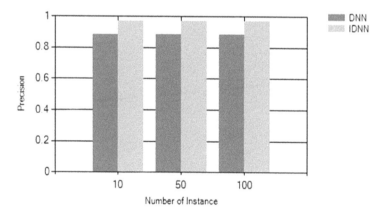

FIGURE 2.8 Performance of DNN vs I_DNN in terms of precision with 1000 train records against 10, 50, 100 test records.

number of testing data records like 10, 50 and 100. For 500 training records with missing data, the obtained recall rate is 94.9 % and with imputed data, the obtained recall rate is 95.8%. Figure 2.10 represents the recall rate analysis of DNN and I_DNN when the proposed ADS system is processed with 1000 training data records against a different number of testing data records like 10, 50 and 100. For 1000 training records with missing data, the obtained recall rate is 87.2 % and with imputed data, the obtained recall rate is 96.9 %.

Figure 2.11 demonstrates the attack detection rate in DNN and I_DNN. From the graph, it is evident that the attack detection rate is more in I_DNN than DNN. Malicious attacks are identified more in I_DNN than DNN. This graph illustrates that in DNN, for 100 instances, nearly 25–30 attacks are detected. But in I_DNN, nearly 30–40 malicious attacks are identified. This graph exhibits that for more number of instances, more attacks could be identified by I_DNN.

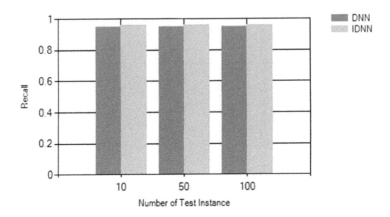

FIGURE 2.9   Performance of DNN vs I_DNN in terms of recall with 500 train records against 10, 50, 100 test records.

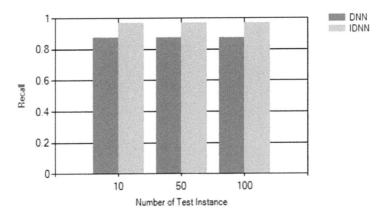

FIGURE 2.10   Performance of DNN vs I_DNN in terms of recall with 1000 train records against 10, 50, 100 test records.

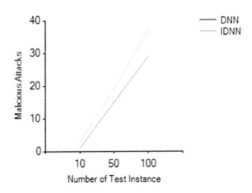

FIGURE 2.11   Detection graph of DNN vs I_DNN.

2.5.4 Performance Comparison of DNN-Based ADS System and EDMDI Algorithm-Based ADS System

Table 2.3 illustrates the performance comparison of the ADS system using the DNN classifier and the proposed EDMDI algorithm-based DNN classifier. The findings in the table demonstrate that the overall performance of the ADS system increases when it implements the proposed algorithm. It can be seen that more malicious attacks are detected for the datasets with 1000 records. So it is evident that the efficiency, accuracy and attack detection rate of the developed ADS model increases when large numbers of records are processed.

From Table 2.1, it can be seen that when 200, 500 and 1000 records are used as training sets in both the existing and proposed systems, the performance of the ADS system increases for the different number of test sets in the proposed system. For instance, when 500 training records are processed with 10 test records in the existing ADS system, the number of malicious attacks detected is 1. But after imputation, when the same numbers of data records are processed in the proposed ADS system, the numbers of malicious attacks detected are 3. Similarly, when 500 training records are processed with 50 test records in the existing ADS system, the numbers of malicious attacks detected are 15. But after imputation, when the same numbers of data records are processed in the proposed ADS system, the numbers of malicious attacks detected are 19. Likewise, when 500 training records are processed with 100 test records in the existing ADS system, the numbers of malicious attacks detected are 29. But after imputation, when the same numbers of data records are processed in the proposed ADS system, the numbers of malicious attacks detected are 37.

For instance, in Table 2.1, when considering the results obtained when 1000 training data records are used, it can be observed that the performance of the proposed ADS system increases with a higher attack detection rate. When 1000 training records are processed with 10 test records in the existing ADS system, the number of malicious attacks detected is 1. But after imputation, when the same numbers of data records are processed in the proposed ADS system, the numbers of malicious attacks detected are 2. Similarly, when 1000 training records are processed with 50 test records in the existing ADS system, the numbers of malicious attacks detected are 13. But after imputation, when the same numbers of data records are processed in the proposed ADS system, the number of malicious attacks detected is 17. Likewise, when 1000 training records are processed with 100 test records in the existing ADS system, the number of malicious attacks detected is 27. But after imputation, when the same numbers of data records are processed in the proposed ADS system, the number of malicious attacks detected is 34.

The experimental results demonstrate that the performance of the ADS system increases when the ADS system is processed with the imputed dataset created by the proposed algorithm. The performance graphs also illustrate that the algorithm performs better for large datasets with many data records; i.e., the performance of the developed ADS system increases when the number of data records used for processing increases. The more the number of records, the more is the significance of missing data imputation.

TABLE 2.3

Performance Comparison of DNN-Based ADS System (with Missing Data) and EDMDI Algorithm with DNN-Based ADS System (with Imputed Data)

Technique	DNN Based ADS System								EDMDI Algorithm with DNN Based ADS System							
No. of Training Records	200		500			1000			200		500			1000		
No. of Testing Records	10	50	10	50	100	10	50	100	10	50	10	50	100	10	50	100
Accuracy	95.9%	95.9%	96.5%	96.5%	89.1%	89.1%	89.1%	96%	96 %	96 %	96.6%	96.6%	96.6%	97.1%	97.1%	97.1%
Precision	94.5%	94.5%	95.6%	95.6%	88.3%	88.3%	88.3%	88.3%	96.2%	96.2%	96.3%	96.3%	96.3%	97.1%	97.1%	97.1%
Recall	93.7%	93.7%	94.9%	94.9%	87.2%	87.2%	87.2%	87.2%	95.8%	95.8%	95.8%	95.8%	95.8%	96.9%	96.9%	96.9%
Time Complexity (sec)	50.37	51.9	145.49	145.60	166.84	345.95	350.59	354.24	34.03	34.55	98.67	97.53	98.14	209.34	211.62	210.43
Malicious Attack Detected	3	19	1	15	29	1	13	27	3	18	3	19	37	2	17	34

2.6 CONCLUSION

A missing data imputation algorithm named EDMDI is proposed. This algorithm fills the missing fields in the dataset with possible values. This algorithm finds the possible value to impute based on the attribute value combinations of the other non-missing records. The value with the highest probability is imputed in the missing field. This algorithm is processed in a newly developed deep learning layer. Then the deep learning technique is used to identify malicious attacks in the processed input. The newly developed ADS is evaluated on different data combinations from the NSL-KDD dataset. The results obtained demonstrate that it identifies more malicious attacks with an increased level of accuracy.

In the future, this work can be expanded for implementation in real-time industrial environments. The input data can be collected from the IIoT environment and processed using the proposed ADS system. An attack warning can be provided in the network if any malicious activity is identified.

REFERENCES

[1] T. Astuti, H. A. Nugroho, T. B. Adji, *"The impact of different fold for cross-validation of missing values imputation method on hepatitis dataset"*, 2015 International Conference on Quality in Research (QiR), 2015.

[2] Robert Mitchell, Ing-Ray Chen, "Behavior rule specification-based intrusion detection for safety critical medical cyber physical systems", *IEEE Transactions on Dependable and Secure Computing*, Vol. 12, Feb. 2015.

[3] Zhun-Ga Liu, Quan Pan, Gregoire Mercier, Jean Dezert, "A new incomplete pattern classification method based on evidential reasoning", *IEEE Transactions on Cybernetics*, Vol. 45, Apr. 2015.

[4] Mohammed A. Ambusaid, Priyadarsi Nanda, "Building an intrusion detection system using a filter-based feature selection algorithm", *IEEE Transactions on Computers*, Vol. 65, No. 10, Oct. 2016.

[5] Hamed Haddad Pajouh, Reza Javidan, Raouf Khayami, Ali Dehghantanha, Kim-Kwang Raymond Choo, "A two-layer dimension reduction and two-tier classification model for anomaly-based intrusion detection in IoT backbone networks", *IEEE Transactions on Emerging Topics in Computing*, Nov. 2016.

[6] Zach DeSmit, Ahmad E. Elhabashy, Lee J. Wells, Jaime A. Camelio, "An approach to cyber-physical vulnerability assessment for intelligent manufacturing systems", *Journal of Manufacturing Systems*, 2017.

[7] Muna AL-Hawawreh, Nour Moustafa, Elena Sitnikova, "Identification of malicious activities in industrial internet of things based on deep learning models", *Journal of Information Security and Applications*, 2018.

[8] Xavier Clotet, Jose Moyano, Gladys Leon, "A real-time anomaly-based IDS for cyber-attack detection at the industrial process level of Critical Infrastructures", *International Journal of Critical Infrastructure Protection*, 2018.

[9] Jose Francisco Colom, David Gil, Higinio Mora, Bruno Volckaert, "Scheduling framework for distributed intrusion detection systems over heterogeneous network architectures", *Journal of Network and Computer Applications*, 2018.

[10] M. Zolanvari, M. A. Teixeira, R. Jain, *"Effect of imbalanced datasets on security of industrial IoT using machine learning,"* 2018 IEEE International Conference on Intelligence and Security Informatics (ISI), 2018.

[11] Amjad Mehmood, Mithun Mukherjee, Syed Hassan Ahmed, Houbing Song, Khalid Mahmood Malik, "NBC-MAIDS: Naive Bayesian classification technique in multi-agent system-enriched IDS for securing IoT against DDoS attacks", *The Journal of Supercomputing*, May 2018.

[12] Muhammad Wasimuddin, Viyaleta Peterson, Karan Manoj Bhosale, Jeongkyu Lee, "*Branching artificial neural ensemble (BRANE): Supervised learning for missing data*", *19th IEEE/ACIS International Conference on Software Engineering, Artificial Intelligence, Networking and Parallel/Distributed Computing (SNPD)*, June 2018.

[13] Berihun Fekade, Taras Maksymyuk, Maryan Kyryk, Minho Jo, "Probabilistic recovery of incomplete sensed data in IoT", *IEEE Internet of Things Journal*, Vol. 5, Aug. 2018.

[14] Cheng-Bo Lu, Ying Mei, "An imputation method for missing data based on an extreme learning machine auto-encoder", *IEEE Access*, Sep. 2018.

[15] Xiang Yu, Xia Fan, Kan Chen, Sirui Duan, "Multi-attribute missing data reconstruction based on adaptive weighted nuclear norm minimization in IoT", *IEEE Access*, Oct. 2018.

[16] Tiago Mazzutti, Mauro Roisenberg, Paulo Jose de Freitas Filho, "*Adaptive missing data imputation with incremental Neuro-Fuzzy Gaussian mixture network (INFGMN)*", *International Joint Conference on Neural Networks (IJCNN)*, Oct. 2018.

[17] Alessio Petrozziello, Ivan Jordanov, Christian Sommeregger, "*Distributed neural networks for missing big data imputation*", *International Joint Conference on Neural Networks (IJCNN)*, Oct. 2018.

[18] Shamsul Huda, Suruz Miah, John Yearwood, Sultan Alyahya, Hmood Al-Dossari, Robin Doss, "A malicious threat detection model for cloud assisted internet of things (CoT) based industrial control system (ICS) networks using deep belief network", *Journal of Parallel and Distributed Computing*, Vol. 120, Oct. 2018.

[19] Ivan Izonin, Natalia Kryvinska, Roman Tkachenko, Khrystyna Zub, "An approach towards missing data recovery within IoT smart system", *Procedia Computer Science*, Vol. 155, 2019.

[20] Daming Li, Lianbing Deng, Minchang Lee, Haoxiang Wang, "IoT data feature extraction and intrusion detection system for smart cities based on deep migration learning", *International Journal of Information Management*, 2019.

[21] J. Kirupakar, S. M. Shalinie, "*Situation aware intrusion detection system design for industrial IoT gateways,*" *2019 International Conference on Computational Intelligence in Data Science (ICCIDS)*, 2019.

[22] M. Zolanvari, M. A. Teixeira, L. Gupta, K. M. Khan, R. Jain, "Machine learning-based network vulnerability analysis of industrial internet of things", *IEEE Internet of Things Journal*, Vol. 6, No. 4, pp. 6822–6834, Aug. 2019.

[23] Mahmudul Hasan, Milon Islam, Ishrak Islam Zarif, M.M.A. Hashem, "Attack and anomaly detection in IoT sensors in IoT sites using machine learning approaches", *Internet of Things*, Vol. 7, Sep. 2019.

[24] Y. Chen, W. Hu, M. Alam, T. Wu, "Fiden: intelligent fingerprint learning for attacker identification in the industrial internet of things", *IEEE Transactions on Industrial Informatics*, 2019.

[25] H. Yao, P. Gao, P. Zhang, J. Wang, C. Jiang, L. Lu, "s", *IEEE Network*, Vol. 33, No. 5, 2019.

3 Internet of Things
Concept, Implementations and Applications

Robina Gujral Bagga and Alok Mishra
Ambalika Institute of Management and Technology,
Uttar Pradesh, India

Kumar Anubhav Tiwari
Kaunas University of Technology, Kaunas, Lithuania

CONTENTS

3.1 INTRODUCTION

The Internet of Things is not only the connection between the computer networks; it's merely the connection between the electronic devices. In the upcoming scenario, we have this point that all the near and surrounding things will be interconnected, such as air conditioners, refrigerators, microwaves, televisions, rice cookers, induction, and any other electronic device. The future is to connect everything and anything as much as possible. So the connection of all the tools, the interconnection between them, and the transfer of data between them is not just the Internet; it's the Internet of Things [7, 11, 16]. Shortly the smart cities and smart homes will be significant examples of the Internet of Things. As all the physical devices will be interconnected, the connection area will be broad so that each device will act as a node for the system. The objective of the Internet of Things is to make things work smartly without involving human interference [25, 32]. As the number of nodes will drastically increase the security concern, data connectivity and data management will be a concern that will be handled accordingly. As far as the Industrial Internet of Things is concerned, it covers machine-to-machine and industrial communication technologies with automation applications [9, 49, 58]. The idea of the Internet of Things is to allow autonomous and secure connections along with an exchange of data between real-world devices and applications. It connects the physical activities of real life with the virtual world. In the present time, the number of connected devices is increasing at a rapid rate. The IoT consists of objects, sensor devices, communication infrastructure, computational and processing units placed on the cloud, and decision-making systems [55, 62, 80, 104]. The idea behind the IoT could also be represented, as shown in Figure 3.1.

For example, suppose a user detects any changes in the refrigerator's temperature with IoT technology. In that case, the user should adjust the temperature with the help of his/her mobile phone [35, 64, 82]. This is a two-way communication to get the work done appropriately (Table 3.1). IoT applications and IoT solutions are widely used in numerous companies across industries (Figure 3.2).

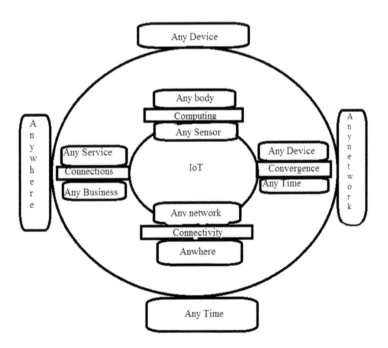

FIGURE 3.1 A's and C's concept in IoT.

TABLE 3.1
Comparison between Traditional Internet and Internet of Things

Topic	Traditional Internet	Internet of Things
Content creator	Human	Machine
Consumption of content	On request	By pushing information and triggering actions
Combination of content	Using defined links	Through explicitly defined operators
What's the value?	Answer questions	Action and timely information
What was done so far?	Both content creation (HTML) and content consumption (search engines)	Mainly content creation

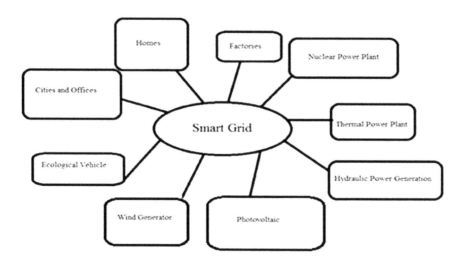

FIGURE 3.2   IoT Applications in various fields.

3.2   LITERATURE REVIEW

Various researchers have contributed to the IoT understanding in multiple fields [97–102]. Researchers have discussed its future possibilities and analysis. The Internet of Things is involved in several areas like connected appliances, smart home security systems, autonomous farming equipment, wearable health monitors, smart factory equipment, wireless inventory trackers, ultra-high-speed wireless Internet, biometric cybersecurity scanners, water level and quality monitoring. According to their description, more than 100 references were analyzed, including articles, books, chapters, practical analysis, etc. Each paper was carefully analyzed and classified into a category. Analysis was performed basically on the initial work done till the present decade. Some of them have been discussed below.

The technologies and ideas that gave rise to the IoT are discussed below [1, 2, 6, 12]:

1832: Baron Schilling created the electromagnetic telegraph in Russia, and in 1833 CarlFriedrich Gauss and Wilhelm Weber invented their code to communicate over a distance of 1200 m within Göttingen, Germany.

1844: Samuel Morse sends the first Morse code public telegraph message "What hath Godwrought?" from Washington, D.C. to Baltimore.

1926: Nikola Tesla in an interview with Collier's magazine, When wireless is perfectly applied, the whole earth will be converted into a huge brain, which it is, all things being particles of a real and rhythmic whole and the instruments through which we shall be able to do this will be amazingly simple compared with our present telephone.

1950: Alan Turing in his article Computing Machinery and Intelligence in the Oxford Mind Journal. It can also be maintained that it is best to provide the machine with the best sense organs that money can buy and then teach it to understand and speak English. This process could follow the normal teaching of a child."

1964: In Understanding Media, Marshall McLuhan stated: "by means of electronic media, we set upa dynamic by which all previous technologies including cities will be translated into information systems."

1966: Karl Steinbuch, a German computer science pioneer said, "In a few decades, computers will be interwoven into almost every industrial product."

1969: Arpanet: The Arpent (an acronym for Advanced Research Projects Agency Network) was the first wide-area packet-switching network with distributed control and one of the first networks to implement the TCP/IP protocol suite. Both technologies became the technical foundation of the Internet. The ARPANET was established by the Advanced Research Projects Agency (ARPA)[90, 92, 96].

1974: Beginnings of TCP/IP

1984: Domain Name System is introduced

1989: Tim Berners-Lee proposes the World Wide Web

1990: Considered the first IoT device, John Romkey created a toaster that could be turned on and off over the Internet for the October '89 INTEROP conference. Dan Lynch, President of Interop promised Romkey that, if Romkey were able to "bring up his toaster on the Net," the appliance would be given star placement in the floor-wide exhibitors at the conference. The toaster was connected to a computer with TCP/IP networking.

1991: Tim Berners-Lee created the first web page

1991: Mark Weiser's Scientific American article on ubiquitous computing called The Computer for the 21st century is written. "The most profound technologies are those that disappear. They weave themselves into the fabric of everyday life until they are indistinguishable from it".

1993: Created by Quentin Stafford-Fraser and Paul Jardetzky the Trojan Room Coffee Pot was located in the "Trojan Room"within the Computer

Laboratory of the University of Cambridge and was used to monitor the pot levels with an image being updated about 3x a minute and sent to the server of the building. It was later put online for viewing once browsers could display images.

1994: Steve Mann creates Wear Cam.

1995: The Internet goes commercial with Amazon and Echo Bay

1997: Paul Saffo's prescient article "Sensors: The Next Wave of Infotech Innovation"

1998: InTouch Scott Brave, Andrew Dahley developed a project at MIT, and Professor Hiroshi Ishii "We then present in Touch, which applies Synchronized Distributed Physical Objects to create a" tangible telephone "for long-distance haptic communication."

1998: A year before losing his battle to cancer Mark Weiser continues his explorations into the topic and constructed a water fountain outside his office whose flow and height mimicked the volume and price trends of the stock market. "Ubiquitous computing is roughly the opposite of virtual reality." Weiser wrote, "Where virtual reality puts people inside a computer-generated world, ubiquitous computing forces the computer to live out here in the world with people".

1999: A big year for the IoT and MIT Linking the new idea of RFID in P&G's supply chain to the then-red-hot topic of the Internet was more than just a good way to get executive attention. It summed up an important insight which is still often misunderstood."

1999: Neil Gershenfeld first spoke about IoT principles in his book "When Things Start to Think."

1999: MIT Auto-ID Lab was initially founded by Kevin Ashton, David Brock, and Sanjay Sharma. They helped to develop the Electronic Product Code

2002: The Ambient Orb created by David Rose and others in a spin-off from the MIT Media Lab is released into the wild with N.Y. Times Magazine named it as one of the Ideas of Year [65, 68, 93].

(2003–2004): RFID is deployed on a massive scale by the U.S. Department of Defense in their Savi program and Wal-Mart in the commercial world.

2005: The U.N.'s International Telecommunications Union (ITU) published its first report on the Internet of Things topic

2008: Recognition by the E.U. and the First European IoT conference is held

2008: A group of companies launched the IPSO Alliance to promote the use of I.P. in networks of "Smart Objects" and to enable the Internet of Things [37, 39].

2008: The FCC voted 5-0 to approve opening the use of the "white space"spectrum (2008–2009): The IoT was born according to Cisco's Business Solutions Group

2008: U.S. National Intelligence Council listed the IoT as one of the six "Disruptive Civil Technologies" with potential impacts on U.S. interests out to 2025.

2010: Chinese Premier Wen Jiabao calls the IoT a key industry for China and has plans to make major investments in Internet of Things.

Some other works done by researchers are as follows:

2010: P. Lalande, J. Bourcier, J. Bardin, and S. Chollet in (2010): This aims at providing architecture and its corresponding runtime to support the creation of self-configuring, self-optimizing and self-repairing applications [94].

2010: Y. Huang and G. Li in 2010: The word thing in the Internet of Things denotes the thing's information. The semantic meaning of the Internet of Things is "an Internet application sharing the items" [45].

2012: Rafiullah, Sarmad Khan, Rifaqat Zaheer, and Shahid Khan in 2012: In this paper, the emerging future of IoT is described as IoT will connect everything[95].

2015: Perumal T., Sulaiman N. Md, Leong C. Y. explained in the paper that water is always a crucial part of everyday life. The level of water is a critical parameter for flood occurrences, especially in disaster-prone areas[66].

2016: Gupta Reetu and Rahul Gupta in 2016: In this paper, the recent developments and challenges that occur in the development of IoT have been defined, and the modern technique useful for device identification has been suggested [72].

2017:A. Pal and B. Purushothaman: explained in their book the security and privacy issue of IoT. The authors have presented a real-world approach to IoT. This book also covers technology components such as communication, computing, storage, and mobility [75].

2017: Priya J, Sailusha Chekuri: In this paper, the authors demonstrated the system that uses containers. The ultrasonic sensors are placed over the boxes to detect the liquid level and compare it with the container's depth [77].

2018: Karwati K, Kustija J in 2018: The authors explained in the paper protocols which describe a vital role in providing convenience in the drainage system. Usually, the water gate at the dam is operated [81].

3.3 SENSORS AND ACTUATORS

This part of the chapter defines various sensors and actuators used in the IoT system. The sensors are devices that detect and respond to changes in environmental situations. The inputs come from a variety of sources such as light, temperature, motion, and pressure. An actuator does the opposite work; it changes an electrical signal to physical action [23, 38, 51, 69]. The sensors come in different shapes and sizes and have their properties. They can be tiny and can be very big. Some sensors can be mechanical; some can be electrical sensors; some may be electronic sensors or chemical sensors. There are so many different types of sensors, and their fabrication is entirely different.

According to the values of different sensors, some action might be required to be taken, and actuators can do that [10, 19, 42].

Some of the necessary sensors are defined below [13, 40, 70, 87]:

(i) The PIR sensor is a passive infrared sensor that can detect any obstacle in the path. So, this is a block-based sensor (Figure 3.3). These sensors are used to see that any object is moved out of the sensor range. They are small in size and low at cost (Figure 3.4).

FIGURE 3.3   Obstacle-based sensor.

FIGURE 3.4   Ultrasonic sensor.

FIGURE 3.5   Camera for Arduino.

These sensors are mostly used in security systems both in domestic as well as in commercial applications.

(ii)   This is an ultrasonic sensor, which detects how far the obstacle is. It has high sensitivity and penetrating power (Figure 3.5).

As the name indicates, ultrasonic sensors measure distance by using ultrasonic waves. The sensor head emits an ultrasonic wave and receives the wave reflected from the target. Ultrasonic Sensors measure the distance to the target by measuring the time between the emission and reception [41, 63, 77, 99].

FIGURE 3.6   Smoke detector.

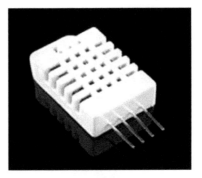

FIGURE 3.7   Temperature and Humidity sensor.

(iii)   Another sensor we have is the camera sensor, which is a small IoT camera. It's an intelligent video camera that can observe, record, analyze, and act accordingly on provided data (Figure 3.6).

(iv)   This is a smoke-detection sensor. So, this sensor can help detect smoke or fire at an early stage, which can help save many lives. This can be very useful to avoid big disasters. If it's enabled with LIVE data, it can send the temperature variation details or smoke information to the fire station directly, which will be helpful in real life [89, 96] (Figure 3.7).

 (v)   Temperature and Humidity sensor: This is the sensor that measures both the humidity and temperature together.

         So, we know that a temperature sensor can measure temperature only. It cannot count, for instance, smoke, or it cannot detect smoke. So, these are very much application-specific [44, 64, 74, 93]. Therefore, DHT11 sensors consist of a NTC (Negative Temperature Coefficient) temperature sensor or a thermistor and an I.C. on the backside of the sensor for measurement.

To measure the humidity-sensing component has two electrodes with moisture-holding substance between them. Therefore as the humidity changes, the resistance between these electrodes changes. Thus for measuring temperature, these sensors use a thermistor. A thermistor is a variable resistor that changes its resistance with temperature change. These sensors are made of ceramic or polymer material to get larger changes in resistance by having just small temperature changes [24, 31, 48, 79].

(vi) Level sensor: A level sensor detects the level of the fluid or fluidized stable. Level sensors can be used as smart waste management for measuring tank levels, diesel fuel gauging, liquid assets inventory, chemical manufacturing high or low-level alarms, and irrigation control. Level sensors can be divided into two groups [14, 29, 61]:

(a) Continuous-level sensors: These can detect the exact position of the fluid. For level detection, usually, proximity sensors, such as ultrasonic or infrared, are used. Capacitive sensors can also be used by recording the changing capacitance value depending on the fluid level. The output can be either analog or digital value.

(b) Point-level sensors: These can detect whether a fluid is above or below the sensor. For the level detection, float or mechanical switch, diaphragm with air pressure or changes in conductivity or capacitance can be used. The output is usually a digital value that indicates HIGH or LOW cost.

Magnetic Reed Liquid Level Sensors: The purpose of a float switch is to open or close a circuit because the level of a liquid rises or falls. Properly used float switches can deliver millions of on/off cycles for years of dependable operation. Failures are generally due to overloading, frequently caused by spiking voltage (Figure 3.8).

Soil Moisture Sensor Module: This is a simple to use digital soil moisture sensor. Just insert the sensor within the soil, measuring moisture or water level content in it. It gives a digital output of 5V when the moisture level is high and 0V when the moisture level is low within the soil [67, 78, 91].

Similarly, actuators are the devices which act on the response of sensors used. They are always behind the performed task. They are not visible during operations.

FIGURE 3.8 Magnetic Reed liquid level sensor.

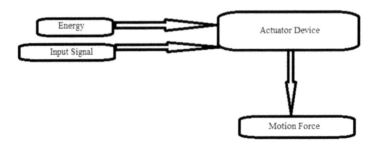

FIGURE 3.9   Actuator working.

They can be separated into four main categories based on their construction pattern and the role they play in a specific IoT environment:

(i) **Linear actuators**– these are used to enable the motion of objects or elements in a straight line.
(ii) **Motors** – these enable precise rotational movements of device components or whole objects.
(iii) **Relays** – this category includes electromagnet-based actuators to operate power switches in lamps, heaters, or even smart vehicles.
(iv) **Solenoids** – these are most widely used in home appliances as part of locking or triggering mechanisms; they also act as controllers in IoT-based gas and water leak monitoring systems (Figure 3.9).

3.4   INTRODUCTION TO ARDUINO

Arduino boards can be divided into six sections depending on their specifications – entry-level, enhanced features, Internet of Things, education, wearable, and 3D printing boards. The most common Arduino boards are *Uno*, *Leonardo*, *Micro*, *Nano* (entry-level), *Mega*, *Pro Mini* (enhanced features). Each of the boards has different specifications and can have various applications [43, 63, 76, 94] (Figure 3.10).

Arduino is an open-source programmable circuit board that can be integrated into a wide variety of complex and straightforward maker space projects. This board contains a microcontroller that can be programmed to sense and control objects in the physical world. By responding to sensors and inputs, the Arduino can interact with a large array of outputs such as LEDs, motors, and displays. Because of its flexibility and low cost, Arduino has become a very popular choice for makers and maker spaces looking to create interactive hardware projects.

There are three different types of memory on the Arduino board: Flash memory, SRAM, and EEPROM. The Flash memory stores the Arduino code, and it is a non-volatile type of memory. That means the information in the memory is not deleted when the power is turned off. The SRAM (static random access memory) is used for storing values of variables when the program of Arduino is running [53, 88]. It has a volatile memory that keeps information only until the power is turned off or the board is reset. The EEPROM (electrically erasable programmable read-only memory) is a non-volatile type of memory used as long-term memory storage.

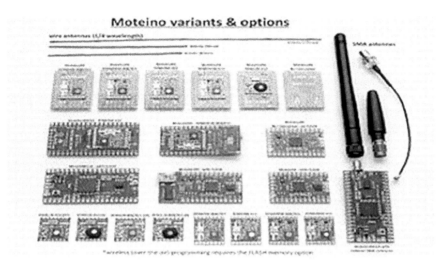

FIGURE 3.10   The most common Arduino boards.

3.4.1   SIZE OF THE BOARD

Arduino microcontrollers have different dimensions of the board, depending on the component located on the board.

3.5   CONCEPT OF SMART CITIES AND SMART HOMES

The idea of smart cities and smart homes is the need of the hour. There are various aspects that we have to focus on accordingly. First, we should know what the meaning of a smart city and smart home is [59, 65].

3.5.1   SMART HOME

This can be defined as the concept that basic household things can be done smartly. Various examples can be given, some we will discuss here:

An example related to the comfort level

A person outside the home will like to control his room temperature before he
    enters, which means he will want to control the A.C. or blower of his room
    from outside before reaching home, which can be done by connecting his
    mobile and the switches in the house.

An example related to safety and security

A person wants to see the interior of the house or each room without physi-
    cally being present, for the purpose because a small child is with a caretaker
    at home or an older person is alone at home, and it will take more time to
    reach home. For safety, they need to watch the cameras from their mobile
    to real-time situations.

An example related to home appliances

A person needs his rice cooker on and starts working so that the hot rice is
ready for him as soon as he/she reaches home, which will help the tired
person relax a bit more. In the same way, the person can switch on or off
the washing machine as needed. Other ways we can say the entire house
switches will be in control of the owner. It will be so comfortable for the
person if they forget to close any appliances in a hurry. He will be able to
switch it off at any time [3, 17, 18, 85, 94].

3.5.2 SMART CITIES

The smart city concept is based on the analysis that various infrastructures should be
connected, including governing bodies. The smart city involves different things, as
follows:

Smart vehicles
Example: A smart connected car with sensors can communicate with a parking
meter and the driver can be directed to the nearest parking spot. The IoT-based
traffic signal can connect with vehicles' sensors, and the analysis of real-time
traffic can be done, and traffic congestion can be controlled. Smart ambulances,
school buses, cars, etc. can let us know if they have any emergency requirement
due to an accident or casualty.

Smart waste management system
Example: Smart garbage cans automatically send data to waste management
companies and arrange to pick up as required rather than a pre-planned pickup.
It will maintain hygiene.

Smart management and governing bodies
Example: The citizen's smartphones will behave like a driver's license and id
card with a digital credential, which speeds the access to the entire city and
governing bodies.

3.6 INTRODUCTION TO IIoT

Industrial IoT (IIoT) can be defined as another form of IoT application. Machine-to-
machine communication and big data analysis and machine learning is a significant
part of IIoT. These data enable the industries to identify and resolve problems with
less effort and more accuracy, which results in overall money and time-saving.
Therefore, IIoT effectively tracks and manages quality control and assurance and
reduces the overall energy consumption [15, 28, 56]. IIoT supports a considerable
amount of industrial data to collect and send to cloud storage and analysis. IIoT can
connect machines, devices, sensors, and people through interconnectivity, helping
companies better address the required demands and safety. IIoT can improve the
maintenance process, overall safety, and connectivity.

3.7   IMPLEMENTATION OF IoT IN VARIOUS FIELDS

The IoT can find its applications in almost every aspect of our daily life. Below are some examples.

Natural disaster prediction: The IoT concept can help in the future forecast of natural calamity in advance by the combination of various sensors and their simulation. These can help to predict the landslide or some other natural disaster and reduce the risk and damage as far as possible [4, 27].

Industry applications: The IoT concept can help manage the fleet of vehicles; they can be cars, school buses, ambulances, etc. The IoT system will help monitor their performance and be connected to know the pickup for the device that needs maintenance [73, 91].

Design of smart homes: The concept of smart homes can be possible with the help of IoT. This concept can help in reducing electricity bills, interaction with appliances, home safety, security, etc.[52, 66].

Medical applications: The IoT can help to find applications in the medical sector for saving lives or improving the quality of life, e.g., monitoring health parameters, tracking activities, support for independent living, monitoring medicines intake, etc.[21, 30].

Agriculture application: IoT concept can be used in various forms in the field of agriculture. The combination of multiple sensors can help monitor the land that needs water or be overfilled with water. The farmer can get the prior intimation of land like pest control, some weather intimations regarding crops, etc. An intelligent way of farming can help the farmers for better growth of crops. It will surely increase productivity [26, 50]. A network of different sensors can sense data, perform data processing, and inform the farmer through communication infrastructure, e.g., a mobile phone text message about the portion of land that needs particular attention. It may include smart packaging of seeds, fertilizer, and pest control mechanisms that respond to specific local conditions and indicate actions. The intelligent farming system will help to have a better understanding of the plant growth models. It will significantly increase agricultural productivity by avoiding inappropriate farming conditions [84, 86].

Groundwater level monitoring: The various sensors available can detect the water-level prediction and the exact status of water supply functional. These multiple sensors can also help in determining the quality of water at that location [5, 22, 33, 47].

Intelligent transport system design: Smart transportation can have many exciting features such as non-stop electronic highway toll, mobile emergency command and scheduling, transportation law enforcement, vehicle rules violation monitoring, reducing environmental pollution, anti-theft system, avoiding traffic jams, reporting traffic incidents, etc.[8, 20, 57].

Design of smart cities: The IoT can help to design intelligent cities, for example, monitoring air quality, efficient lighting up to the town, watering gardens, intelligent buses, smart government organizations, traffic control, and automatic vehicle monitoring, etc.[83, 90].

Smart Security: Smart security will be the essential factor in IoT management. As we all know, safety and security are very necessary in real life. As the population is increasing, so the data consumption for each and every task will increase rapidly. The IoT applications will continuously evolve much information, but it has to face various challenges related to privacy, security, and complexity. So, smart management and security systems will be the desire of the future, as all the things will be smartly connected [34, 46].

3.8   CHALLENGES IN IoT

Challenges in the field of IoT are anonymous; some of them have been discussed below:

3.8.1   PRIVACY CONCERN

Privacy is the biggest issue nowadays. Presently, every device is connected via the Internet, which increases the risk of data loss. Moreover, it may not be secure and easily damaged by the third party when we exchange any confidential information.

3.8.2   COMPLEXITY

Along with significant advantages, the system's complexity will also increase; for example, if some system error occurred without our notice, the wrong message can be sent anywhere, it may be a hospital, grocer, or office anywhere the incorrect information can be sent [60, 71].

3.8.3   DEPENDENCY

As far as dependency is concerned, that's true as human life has become so fast with the help of technology that we cannot expect to continue without them. From morning prayers on Alexa to evening dimming the light by Google home, our daily life is affected.

3.8.4   SOCIETY

The consumer's demands will increase with time; along with this, understanding the needs will also change as the requirement will adjust accordingly. As everything will be interlinked, all the actions will happen quickly; before the previous problem is resolved, a new issue will be solved [36, 54].

3.9   CONCLUSION AND FUTURE ASPECT

So, finally, we can conclude that the IoT field is vast enough. It is involved in every field of life. Various sensors, actuators, Arduino boards, etc., can jointly create many practical applications that can be included in our day-to-day life, making our lives easier and more comfortable.

The future of IoT will have each and everything connected. Life will have a comfortable level and from the security point of view, we have to be alert as the most robust connectivity and security will be required to fulfill all the desired needs. Especially for the smart cities implementation, the government bodies will be involved; for the smooth working, security, and safety will be the priority. In the future, all the fields like agriculture, medical, defense, governing bodies, education, etc., will all be IoT-based.

REFERENCES

[1] P.B. Jones, G.D. Walker, and R.W. Harden, L.L. McDaniels, "The Development of Science of Hydrology," Circular No 63-03, Texas Water Commission, April 1963.

[2] W.C. Palmer, "Meteorological Drought," Research Paper No. 45, U.S. Weather Bureau, Washington, DC. 1965.

[3] D.A. Wilhite, N.J. Rosenburg, and M.H. Glantz, "Improving Federal Response to Drought," *Journal of Climate and Applied Meteorology*, 25(3), pp. 332–342, 1986.

[4] T.B. McKee, N.J. Doeskin, and J. Kleist, "*The Relationship of Drought Frequency and Duration to Time Scales,*" *Eighth Conference on Applied Climatology*, American Meteorological Society, Boston, 1993.

[5] T.B. McKee, N.J. D oeskin, and J. Kleist, "*Drought Monitoring with Multiple Time Scales,*" *Ninth Conference on Applied Climatology*, American Meteorological Society, Boston, MA, 1995.

[6] D.B. Johnson and D.A. Maltz, "Dynamic Source Routing in Ad Hoc Wireless Networks," *Mobile Computing*, Springer: New York, NY, 1996, pp. 153–181.

[7] D.A. Wilhite, "State Actions to Mitigate Drought: Lessons Learned," *Journal of the American Water Resources Association*, 33(5), pp. 961–968, 1997.

[8] J. Broch, D.A. Maltz, D.B. Johnson, Y.-C. Hu, and J. Jetcheva, "*A Performance Comparison of Multi-Hop Wireless Ad Hoc Network Routing Protocols,*" *Proceedings of the 4th Annual ACM/IEEE International Conference on Mobile Computing and Networking (ACM)*, Dallas, TX, October 25–30, 1998, pp. 85–97.

[9] W.B. Solley and R.R. Pierce, "Estimated Use of Water in the United States in 1995," US. Geological Survey Circular, 1200, 1998.

[10] M.J. Neilson and D.N. Bearce, "Seasonal Variations in Water Table Elevations in the Surficial Aquifer, Birmingham Valley," *Journal of the Alabama Academy of Science* 69(3), pp. 175–182, 1998.

[11] H.R. Byun and D.A. Wilhite, "Objective Quantification of Drought Severity and Duration," *Journal of Climate* 12(2), pp. 742–756, 1999.

[12] J.A. Butterworth, R.E. Schulze, L.P. Simmonds, P. Moriarty, and F. Mugabe, "Long-Term Groundwater Level Fluctuations due to Variation in Rainfall," *Hydrology and Earth System Sciences* 3(3), 1999.

[13] M. Weiser, R. Gold, and J.S. Brown, "The Origins of Ubiquitous Computing Research at PARC in the Late 1980s," *IBM Systems Journal* 38(4), pp. 693–696, December 1999.

[14] E.U. Water Framework Directive (Directive 2000/60/E.C.), 2000.

[15] Microcontroller Chip Technology, 2001, PIC16F84A Datasheet, www.microchip.com.

[16] C.J. Taylor and W.M. Alley, "Ground-Water-Level Monitoring and the Importance of Long-Term Water-Level Data," U.S. Geological Survey Circular 1217, 2001.

[17] W.M. Wendland, "Temporal Responses of Surface-Water and Ground-Water to Precipitation in Illinois," *Journal of the American Water Resources Association* 37(3), pp. 685–693, 2001.

[18] W.W. Brent, G.F. Paul, T. Sisk, and W.K. George, *"Wireless Sensor Networks For Dense Spatio-Temporal Monitoring of The Environment: A Case For Integrated Circuit, System, And Network Design,"* Proc. *2001 IEEE CAS Workshop on Wireless Communications and Networking*, August 2001.

[19] B. Warneke, M. Last, B. Liebowitz, and K. Pister, "Smart Dust: Communicating with a Cubic-Millimetre Computer," *IEEE Computer Magazine* 34(1), 44–51, January 2001.

[20] D. Braginsky and D. Estrin, *"Rumor Routing Algorithm for Sensor Networks,"* *Proceedings of the First ACM International Workshop on Wireless Sensor Networks & Applications*, Atlanta, GA, 28 , pp. 22–31, September 2002.

[21] D.A. Wilhite, "Combating Drought through Preparedness," *Natural Resources Forum* 26(4), pp. 275–285, 2002.

[22] Tearfund, "Water," http://www.tearfund.org/uploads/documents/water(1).pdf, 2002.

[23] W. Ye, J. Heidemann, and D. Estrin, "An Energy-Efficient MAC Protocol for Wireless Sensor Networks," *Proceedings of IEEE INFOCOM*, 3, 1567–1576 June 2002.

[24] AENOR EN 61512: Batch Control – Part 2: Data Structures and Guidelines for Languages, 2002.

[25] P. Kinney, *"Zigbee Technology: Wireless Control That Simply Works,"* *Proceedings of the Communications Design Conference*, San Jose, CA, 29 September–2 October 2003, pp. 1–7.

[26] J. Luo, P.T. Eugster, and J.P. Hubaux, *"Route Driven Gossip: Probabilistic Reliable Multicastin Ad Hoc Networks,"* *Proceedings of the Joint Conference of the CiteSeer IEEE Computer and Communications*, San Francisco, CA, 30 March–3 April 2003, 3, pp. 2229–2239.

[27] J. Imtiaz and J. Jasperneite, *"Scalability of OPC-UA Down to the Chip Level Enable the Internet of Things,"* *Proc. of the 11th IEEE International Conference on Industrial Informatics (INDIN'13)*, Bochum, Germany. IEEE, July 2013, pp. 500–505. 20 PSI Directive (2003/98/E.C.). Directive on the Re-Use of Public Sector Information Entered into Force on 31 December 2003.

[28] D. Misiones, J.P. Vitkovskyt, G. Olsson, A.R. Simpson, and M.F. Lambert, "Pipeline Burst Detection and Location Using a Continuous Monitoring Technique," *Proc. Intl. Conf. on Computing and Control for the Water Industry (CCWI)*, 89–96, 2003.

[29] S. Anumalla and D.C. Gosselin, "Ground Water Monitoring Using Smart Sensors," https://www.researchgate.net/publication/241599056, January 2004.

[30] http://www.pldworld.com/actel/html/ref/glossary-security-body.htm, accessed on June 3 2004, http://en.wikipedia.org/wiki/802.11b, accessed on *3 June 2004*.

[31] G. Lu, B. Krishnamachari, and C. Raghavendra, *"An Adaptive Energy-Efficient and Low Latency MAC for Data Gathering in Sensor Networks,"* *IEEE WMAN, 0-7695-2132-0/04/ Proceedings of the 18th International Parallel and Distributed Processing Symposium (IPDPS'04)* April 2004.

[32] A. Baggio, *"Wireless Sensor Networks in Precision Agriculture,"* *Proceedings of the ACM Workshop on Real-World Wireless Sensor Networks (REALWSN 2005)*, Stockholm, Sweden, 20–21 June 2005.

[33] D. Misiunas, M. Lambert, A. Simpson, and G. Olsson, "Burst Detection and Location in Water Distribution Networks," *Water Science and Technology: Water Supply*, 5(3–4), 71–80, 2005.

[34] Y. Zhou, Z. Ling, and Q. Wu, "Zig Bee Wireless Communication Technology and Investigation on Its Application," *Process Autom. Instrum.* 2005.

[35] N. Wang, N. Zhang, and M. Wang, "Wireless Sensors in Agriculture and Food Industry Recent Development and Future Perspective," *Comput. Electr. Agric.*, 50, pp. 1–14, 2006.

[36] E.J. Cho and F.V. Bright, "Integrated Chemical Sensor Array Platform Based on the Light Emitting Diode, Xerogel Derived Sensor Elements, and High-Speed Pin Printing," *Analytica Chimica Acta*, 470, pp. 101–110, 2007.

[37] D. Ganesan, B. Krishnamachari, A. Woo, D. Culler, D. Estrin, and S. Wicker, "Complex Behaviour at Scale: An Experimental Study of Low-Power Wireless Sensor Networks," UCLA CS Technical Report UCLA/CSD-TR020013, 2008.

[38] A. Bergant, A.S. Tusseling, J.P. Vitkovsky, D.I.C. Covas, A.R. Simpson, and M.F. Lambert, "Parameters affecting water-hammer wave attenuation, shape and timing –Part 1: mathematical tools," *Journal of Hydraulic Research*, 46(3), pp. 373–381, 2008.

[39] J. Hill and D. Culler, "Mica: A Wireless Platform for Deeply Embedded Networks," *IEEE Micro*, 22(6), pp. 12–24, November–December 2008.

[40] J. Zhang, S. Hu, Z. Long, and Q. Kou, "The Wireless Data Transmission System Based on GPRS and Its Discussion for Application," *J. Electr. Meas. Instrum.*, 23, p. S1, 2009.

[41] M. Javanmard, K.A. Abbas, and F. Arvin, "A Microcontroller-Based Monitoring System for Batch Tea Dryer, CCSE, "*Journal of Agricultural Science*, 1(2), December 2009.

[42] P. Lalanda, J. Bourcier, J. Bardin, and S. Chollet, *Smart Home Systems, Smart Home Systems*, Mahmoud A. Al-Qutayri (Ed.), ISBN: 978-953-307-050-6, InTech, 2010, Available from: http://www.intechopen.com/books/smart-home-systems/smart-home-systems.

[43] L. Tan and N. Wang, "*Future Internet: The Internet of Things*," *Proceedings of the IEEE 3rd International Conference on Advanced Computer Theory and Engineering (ICACTE)*, Chengdu, China, 20–22 August 2010.

[44] L. Atzori, A. Iera, and G. Morabito, "The Internet of Things: A survey," *Computer Networks*, 54(15), pp. 2787–2805, October 2010.

[45] Y. Huang and G. Li, "*Descriptive Models for Internet of Things*," *IEEE International Conference on Intelligent Control and Information Processing (ICICIP)*, August 2010.

[46] M. Allen, A. Preis, M. Iqbal, S. Srirangarajan, H.B. Lim, L. Girod, and A.J. Whittle, "Real-Time In-Network Distribution System Monitoring to Improve Operational Efficiency," *Journal American Water Works Association (JAWWA)*, 103(7), pp. 63–75, 2011.

[47] H. Kopetz. Internet of Things. In *Real-Time Systems*; Springer: New York, NY, USA, 2011, pp. 307–323.

[48] Y.H. Lin, *Introduction to the Internet of Things*, Science Press: Beijing, China, pp. 60–63, 2011.

[49] D. Miorandi, S. Sicari, F.D. Pellegrini, and I. Chlamtac, "Internet of Things: Vision, Applications and Research Challenges," *Ad Hoc Networks*, 10(7), pp. 1497–1516, September 2012.

[50] K. Rafiullah, et al., "*Future Internet: The Internet of Things Architecture, Possible Applications and Key Challenges*," *2012 10th International Conference on Frontiers of Information Technology*, 2012.

[51] L. Perelman, J. Arad, M. Housh, and A. Ostfeld, "Event Detection in Water Distribution Systems from Multivariate Water Quality Time Series," *Environmental Science and Technology, ACS*, 46, pp. 8212–8219, 2012.

[52] D.L. Li, "Internet of Things and Wisdom of Agriculture," *Agric. Eng.*, 2, p. S126, 2012.

[53] B. Tian, X.L. Zhao, Q.M. Yao, and L. Zha, "*Design and Implementation of a Wireless Video Sensor Network*," *Proceedings of the 2012 9th IEEE International Conference on Networking, Sensing and Control (ICNSC)*, Beijing, China,11–14 April 2012, pp. 411–416.

[54] R. Alcarria, T. Robles, A. Morales, D.L. deIpina, and U. Aguilera, "Enabling Flexible and Continuous Capability Invocation in Mobile Prosumer Environments," *Sensors*, 12(7), pp. 8930–8954, June 2012.

[55] Y.J. Fei, Z.J. Xu, and L. Feng, *"The Research of Internet of Things in Agricultural Production and Management,"* Proceedings of the Fifteenth Session of the Annual Meeting of the Association of China, the Tenth Venue: Conference on Information Technology and Agricultural Modernization, Guiyang, China, 25–27 May 2013.

[56] K. Thorin, "Know the Difference between the Most Common Arduino Boards," *Lifehacker,* 13 June 2013, lifehacker.com/know-the-difference-betweenthe-most-common-Arduino-boa-513094593

[57] W. Zhang, "Research on Key Technologies of Wireless Sensor Networks for Precision Agriculture," PhD. Thesis, Zhejiang University, Hangzhou, China, 2013.

[58] M. Lee, J. Hwang, and H. Yoe. *"Agricultural Production System Based on IoT,"* Proceedings of the 2013 IEEE 16th International Conference on Computational Science and Engineering (CSE), Sydney, Australia, 3–5 December 2013, pp. 833–837.

[59] J. Gubbi, R. Buyya, S. Marusic, and M. Palaniswami, "Building the Common Future We Want" United Nations Department of Economic and Social Affairs, 2013.

[60] P. Kumar, S. SPathan, and B. Mashilkar, "Liquid Level Control using PID Controller Based on Labview & Matlab Software," *International Journal of Engineering Research & Technology (IJERT),* 3(10), pp. 111–114, October 2014.

[61] United Nations, "Global Sustainable Development Report–Executive Summary: Building the Common Future We Want," United Nations Department of Economic and Social Affairs, Division for Sustainable Development: New York, 2013.

[62] T. Robles, R. Alcarria, D. Mart'ın, and A. Morales, *"An Internet of Things-Based Model for Smart Water Management,"* Proc. of the 8th International Conference on Advanced Information Networking and Applications Workshops (WAINA'14), Victoria, Canada. IEEE, May 2014, pp. 821–826.

[63] R. Alcarria, T. Robles, A. Morales, and E. Cedeno, "Resolving Coordination Challenges in Distributed Mobile Service Executions," *International Journal of Web and Grid Services,* 10(2), pp. 168–191, January 2014.

[64] E. Serrano, P. Moncada, M. Garijo, C. Iglesias, "Evaluating Social Choice Techniques into Intelligent Environments by Agent-Based Social Simulation," *Information Sciences,* 286, pp. 102–124, December 2014.

[65] M.R. Cook, N.E. Moss, and M. Dorina, "Analysis of Sediment Loading Rates for the Magnolia River Watershed, Baldwin County, Alabama, 2009," Geological Survey of Alabama Open-file Report 1409, 22 p, 2014.

[66] T. Perumal, N. Sulaiman, and C.Y. Leong, *"Internet of Things (IoT) Enabled Water Monitoring System,"* IEEE 4th Global Conference on Consumer Electronics, pp. 86–87, 2015.

[67] L. Dan, C. Xin, C. Huang, and L. Ji, *"Intelligent Agriculture Greenhouse Environment Monitoring System Based on IoT Technology,"* Proceedings of the 2015 IEEE International Conference on Intelligent Transportation, Big Data and Smart City (ICITBS), Halong Bay, Vietnam, 19–20 December 2015, pp. 487–490.

[68] S. Madakam, R. Ramaswamy, and S. Tripathi, "Internet of Things (IoT): A Literature Review," *Journal of Computer and Communications,* 3(5), 2015.

[69] F.J. Ferrández-Pastor, J.M. García-Chamizo, M. Nieto-Hidalgo, J. Mora-Pascual, and J. Mora Martínez, "Developing Ubiquitous Sensor Network Platform Using Internet of Things: Application in Precision Agriculture," *Sensors,* 16, p. 1141, 2016.

[70 S. Zhao, L. Yu, and B. Cheng, "A Real-Time Web of Things Framework with Customizable Openness Considering Legacy Devices," *Sensors,* 16, p. 1596, 2016.

[71] S.C. Abraham, "Internet of Things (IoT) with Cloud Computing and Machine-to Machine (M2M) Communication," *International Journal of Emerging Trends in Science and Technology,*3, 4654–4661, 2016.

[72] R. Gupta and R. Gupta, *"ABC of Internet of Things: Advancements, Benefits, Challenges, Enablers and Facilities of IoT,"* 2016 Symposium on Colossal Data Analysis and Networking (CDAN), pp. 1–5, 2016. doi: 10.1109/CDAN.2016.7570875.

[73] J. Wenwen, Z. Linbo, Z. Feifan, G. Wenjing, and G. Yuxin,"Intelligent Wireless Environmental Monitoring System of University Laboratory Based on the Internet of Things," in *Internet of Things (IoT) and Engineering Applications*, Clausius Scientific Press, Canada, 2016, ISSN 2371-8617 (Online).

[74] Blog, "Telephone System Engineer,"18 May 2017, telephonesystemspro.co.uk/blog.

[75] A. Pal and B. Purushothaman, *IoT Technical Challenges and Solutions*. Artech House, 2017.

[76] G. Palavicini, Jr, J. Bryan, E. Sheets, M. Kline, and J. S. Miguel, *"Towards Firmware Analysis of Industrial Internet of Things (IIoT) - Applying Symbolic Analysis to IIoT Firmware Vetting,"* Proceedings of the 2nd International Conference on Internet of Things, Big Data and Security, 2017.

[77] J. Priya and S. Chekuri, "Water Level Monitoring System Using IoT," *International Research Journal of Engineering and Technology (IRJET)*, 4(12), pp. 1813–1817, December .2017.

[78] T. Yang, B. Di Martino, and Q. Zhang, "Internet of Everything," *Mobile Information Systems*, 2017, 2017.

[79] "IoT: A Hot Topic for the Future," Pepeeta, Keepmemag. "Future Internet," Telephone System Engineer, 1 May 2017, www.pepeeta.com/internet_of_things/.

[80] F. Hassan, A.R. Khan, and S. A. Madani, *Internet of Things: Challenges, Advances, and Applications*, Chapman & Hall/CRC Computer and Information Science Series, CRC Press, 2017.

[81] K. Karwati and J. Kustija, "Prototype of Water Level Control System," *IOP Conference Series: Materials Science and Engineering*, 384, 2018, 012032. doi: 10.1088/1757-899X/384/1/012032.

[82] A. Chaudhuri, "Internet of Things and Its Potential," *Internet of Things, for Things, and by Things*, pp. 3–16, 2018.

[83] Seebo Blog, "Ultrasonic Sensors: Applications for the Internet of Things," 9 July 2018, blog.seebo.com/IoT-ultrasonic-sensors/.

[84] T. Shetty, P. Wagh, and A. Dudwadkar, "Water Level Monitoring System," *International Research Journal of Engineering and Technology (IRJET)*, 5(8), pp. 1712–1714, August 2018.

[85] V. Jeevagan and S. Prem Kumar, "Water Level Monitoring System Using IoT&ATmega 328p Microcontroller," *International Journal of Pure and Applied Mathematics*, 119(18), 1497–1501, 2018.

[86] M. Noura, M. Atiquzzaman, and M. Gaedke, "Interoperability in Internet of Things: Taxonomies and Open Challenges," *Mobile Networks and Applications*, 24(3), pp. 796–809, 2018.

[87] Segun O. Olatinwo and Trudi-H. Joubert, Senior Member, IEEE, "Enabling Communication Networks for Water Quality Monitoring Applications: A Survey," Febuary 2019.

[88] Murray D. Einarson, John A. Cherry, "A New Multilevel Groundwater Monitoring System Using Multichannel Tubing," NGWA.

[89] Arduino Getting Started, "Arduino-Temperature Humidity Sensor: Arduino Tutorial," arduinogetstarted.com/tutorials/arduino-temperature-humidity-sensor.

[90] Arduino Tech, "Arduino Boards," www.arduino-tech.com/arduino-boards/.

[91] KEYENCE, "Detection Based on 'Ultrasonic Waves': What Is an Ultrasonic Sensor?" www.keyence.com/ss/products/sensor/sensorbasics/ultrasonic/info/.

[92] United States Department of Defence, https://en.wikipedia.org/wiki/United_States_Department_of_Defense

[93] The New York Times Magazine, https://www.nytimes.com/2002/12/15/magazine/the-year-in-ideas-news-that-glows.html.

[94] P. Lalanda, J. Bourcier, J. Bardin, and S. Chollet, *Smart Home Systems, Smart Home Systems*, Mahmoud A. Al-Qutayri (ed.), Intech Open, 1 February 2010, doi: 10.5772/8415. Available from: https://www.intechopen.com/books/smart-home-systems/smart-home-systems.

[95] R. Khan, S.U. Khan, R. Zaheer, and S. Khan, *"Future Internet: The Internet of Things Architecture, Possible Applications and Key Challenges,"* Conference: *Frontiers of Information Technology (FIT)*, 2012.

[96] T. Perumal, *Internet of Things (IoT) Enabled Water Monitoring Systems. Conference: 2015 IEEE 4th Global Conference on Consumer Electronics (GCCE)* Osaka, Japan 2015. doi: 10.1109/GCCE.2015.7398710.

[97] A. Solanki and A. Nayyar, "Green Internet of Things (G-IoT): ICT Technologies, Principles, Applications, Projects, and Challenges," *Handbook of Research on Big Data and the IoT*, pp.379–405, IGI Global, 2019.

[98] R. Krishnamurthi, A. Nayyar, and A. Solanki, "Innovation Opportunities through Internet of Things (IoT) for Smart Cities," *Green and Smart Technologies for Smart Cities*, CRC Press, Boca Raton, FL, pp. 261–292, 2019.

[99] P.K.D. Pramanik, A. Solanki, A. Debnath, A. Nayyar, S. El-Sappagh, and K.S. Kwak, "Advancing Modern Healthcare with Nanotechnology, Nanobiosensors, and Internet of Nano Things: Taxonomies, Applications, Architecture, and Challenges," *IEEE Access*, 8, pp.65230–65266, 2020.

[100] A. Nayyar, R. Rameshwar, and A. Solanki, "Internet of Things (IoT) and the Digital Business Environment: A Standpoint Inclusive Cyber Space, Cyber Crimes, and Cybersecurity," *The Evaluation of Business in Cyber Age*, Apple Academic Press Inc., pp. 111–151, 2020.

[101] H. Kaur, S.P. Singh, S. Bhatnagar, and A. Solanki, "Intelligent Smart Home Energy Efficiency Model Using Artificial Intelligence and Internet of Things," *Artificial Intelligence to Solve Pervasive Internet of Things Issues*, pp. 183–210, Academic Press, 2021.

[102] S.P. Singh, A. Solanki, T. Singh, and A. Tayal, "Internet of Intelligent Things: Injection of Intelligence into IoT Devices," *Artificial Intelligence to Solve Pervasive Internet of Things Issues*, pp. 85–102, Academic Press, 2021.

4 Output-Oriented Multi-Pane Mail Booster

Data Crawling and Results in All Category Panes of a Mail

S. Hrushikesava Raju, V. Lakshmi Lalitha,
Praveen Tumuluru, N. Sunanda, and S. Kavitha
Koneru Lakshmaiah Education Foundation, Vaddeswaram,
Guntur

Saiyed Faiayaz Waris
Vignan's foundation for science, Technology and Research,
Vadlamudi, Guntur

CONTENTS

4.1 INTRODUCTION

The usage of mails is nowadays becoming a normal routine, but after a long time, usage leads to many mails. The normal user may be considered where the user has to search for a mail with attachments whose content is somewhere matched with the pattern taken for search. The factors considered in this scenario are customer satisfaction, readability, and efficient searching. The scenario takes the input as a pattern, it searches for not only the mails but also whose attachment content matched with the input pattern. Normally, the group of conversations be returned only when searching for the input given. In the proposed scenario, the attachments are considered a study where a mail text is searched against mail attachments. The searching is limited to the current scope that may be inbox or sent or spam or promotions or other category panes. This searching does not apply to all panes at a time. This proposed work focuses on all panes for the attachments. The normal approach is on mails content along with the objective of intended scope. This could be focused on the proposed approach. The disadvantages identified are taking more time to search for a certain pattern in the mails and generating statistical analysis over the searching.

The user can make customized folders and store their content in such folders for easy navigation. The folders are here of two types: pre-defined folders and user-defined folders. The user-defined folders elaborate the theme where the user would store the specific mails in such customized folders against other normal mails for quick identification. But here, remembering the content of all mails in the specific folders is again a time-consuming process. Hence, the seed idea is that the input pattern is for searching the content of usual box mails and the attachments. The attachments could be of many types available such as docx, pdf, txt, rtf, doc, and other text-based types.

The steps involved are to be defined in order to produce the expected result of the intended theme. The intended task is to search for a mail content as the pre-defined query search engine that considers text-type attachments also. The observation is taken on text-related file types such as doc, docx, rtf, pdf, txt, and other related types. The output produced a statistical report which describes whatever panes in which the input pattern is matched are brought on the fore-end. The user can now easily navigate through any pane that possesses mails with matched content as well as attachments are found. The user has more customer satisfaction in this proposed approach.

There are a variety of file types that may be attached, but text-oriented files are attached most of the times. But the extended scope of work that includes the attachments like audio or video in addition to the text.

The expected work considers the below steps:

1. Provide a pattern in the search option available in any authorized mail provider.
2. Among the many searchable algorithm types, the KMP is taken because of fewer mismatched comparisons.
3. Once the pattern is found in the attached files, those mails are also to be returned in the output mails.
4. Not only the inbox should be checked but checking applicable to all the panes.
5. Finally, evaluating the proposed approach functionality is based on the significant factors.

4.2 RELATED WORK

There are no studies found exactly on the intended theme but the most related studies on the problem identified are taken as study. The demonstration of works in [1–3] denotes the useful descriptions on challenges and safety precautions are to be taken over the issues raised, and some methods are also listed to find out the cause of hacking the mails in terms of passwords breaking, phishing, etc. As per orientation of sources mentioned in [4–6], the benefits are listed like free service, speed in transmitting the content, no overhead cost, etc. as well as demerits listed such as downloading files along with viruses, always need of internet, phishing, etc.

The mail usage possesses certain benefits as well as pitfalls and is listed below in a diagram (Figure 4.1).

The work demonstration in [7] states responses regarding questions and online strata from the organization's employees. The description mentioned in [8] denotes

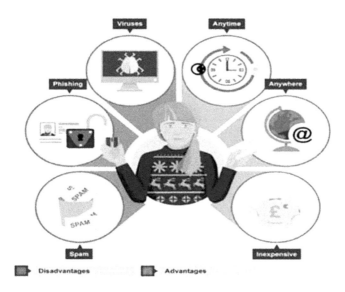

FIGURE 4.1   Emails – benefits and drawbacks.

the process of making decisions based on social media and the use of their content in giving the right information. Regarding the information provided in [9], the cost incurred for alerting in the view of the receiver's aspect and benefits and pitfalls is listed out based on the discussion made. The information mentioned in [10–12] describes what actions should be taken for the obtained content when certain text is searched in the specific mail service providers. With respect to the source mentioned in [13], the clustering technique is proposed to analyze tweeted text. The work elaborated in the source mentioned in [14] states that retrieving information is based on different management systems such as tourism and all types of leave. As per the direction of the text mentioned in [15], purchasing impact of a specific city Vijayawada and extracting the client's opinions are demonstrated with suitable tools.

According to [16], commerce e-service sites with respect to the true opinions lead to the extraction of beliefs and trust. With regard to [17], the focus is on spam and the identification of mails in the spam using the semantic web. In [18], the application of context similarity in the mails and their impact through the text mode is described. As per [19], the application of similarity scale for text identicalness and their impact in the text modes are described. With respect to the description from [20, 21], the two kinds of similarity searches such as context and semantic are used over the text that is floated in the websites for identicalness and note their impact. With respect to [22], OTP approach is involved in DNA-based cryptography to provide security. As per [23], security is guaranteed using the available accessing models and encryption techniques. The source mentioned in [24] states how a network is secured using IoT and what features made it possible. As per [25], the challenges in IoT in terms of security and privacy are listed and are focused on in an elaborated manner (Figure 4.2).

Regarding [26], IoT is used in such way that automation of testing the eyes and ordering the spectacles is done and their order of activities is mentioned. As per [27],

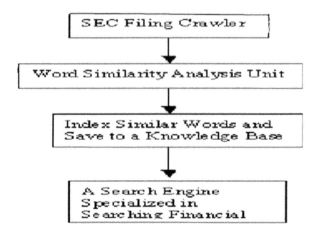

FIGURE 4.2   Searching approach and its steps.

the infected surroundings are detected spontaneously using efficient sensors under IoT. In the view of [28], sensors are activated to load the current region and apply that country's policies in judging the currency of that region, update the accounts after payment. Regarding [29], an app is developed that detects the gadgets that would transfer the requested charging such as Phone Pe. In the aspect of [30], the detection of weighted objects that fall from the top of high buildings, catch them and that would deposit at the ground safely using IoT and its sensors.

Regarding [31], IoT is used to detect the leakages in advance and rectify the scenarios from dangers using IoT. In the view of [32], sensors and IoT are used in determining the components that would give energy and taste, provide statistics of health in the future, and provide precautions. In [33], IoT and Maps are used in determining the ranking of the places and recommend to the users who would plan their journeys effectively—concerning [34], increasing the crops by monitoring the agriculture using IoT and related sensors based on disease prediction as well as weather prediction. As per [35], IoT is used in automating the activities of a house and minimize human efforts and provide security to be ensured. According to [36], video surveillance is used in which specific activities are tracked using color codes at the moving bar of the video to identify easily. As per [37], IoT and satellite technology are used in driving over the roads and providing safe alerts about the road conditions. With regard to [38], the specific defined method is applied in determining the specific patterns over certain files and producing the statistics of results at a time.

Every study mentioned here is useful in searching for identicalness and close plagiarism and is somehow helpful in guiding the intended theme, but is not exactly what the intended task aims for.

4.3   PROPOSED METHODOLOGY

Here, the intended theme is divided into the modules where searching in a usual mode is the traditional approach and the other is searching over the mails that consist of attachments.

a. **Usual searching:** Normal searching where mails' content matched with the input pattern
b. **Searching on mails with attachments:** Multiple panes on which searching is applied over mails having attachments and returning the matched mails in the multiple panes if matching found in panes.

The proposed approach takes these pieces into three modules for convenience. The intended task work is specified in the following algorithm:

ALGORITHM STATISTICS_OF_TEXT_OVER_MAILS (MAIL_SERVICE)

Step 1: Call login module where login (user, pass, OTP, certain_text)

    1.1 After verifying the details of the user, verify the more levels of authentication for security.
    1.2 Once all details are verified, access is granted.

The functionality of mail service module is depicted as in Figure 4.3:

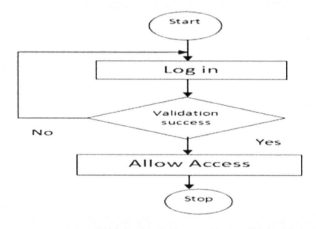

FIGURE 4.3 Login authentication functionality.

Step 2: Call similarity (certain_text):

    2.1 cmail=0
        For email that has attachment:
            attachment converted into text file using available online converters
            cmail=cmail+1
    2.2 For finding the similarity of certain_text over the text file, KMP is applied for efficiency
        Pseudo_Procedure KMP (pattern,file_content):
            2.2.1 Generate failure function table based on principle – the longest prefix of pattern is the length for each index in the pattern.

2.2.2 When traversing the pattern against text, m and n are lengths of pattern and texts.

```
while(j>0 && Y[j] != Y[i]):
        j=next[i]
if (j > 0 || Y[j] == Y[i])
        next[i + 1] = j + 1;
```

2.3 If the certain_text has not matched with mail with attachment
cmail=cmail-1
2.4 Apply this knowledge not only to the current context but also to other category panes.
2.5 Display the mails with respect to panes in each separate dialog box as per customer need.

The functionality of searching against the mails with respect to the category panes is demonstrated in Figure 4.4:

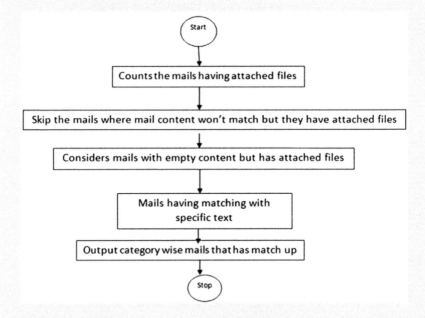

FIGURE 4.4   Searching the input pattern in the proposed approach.

Step 3: Statistics module is called where the activities are listed out

3.1 panes are counted in terms of cpane=0
cpane=cpane+1
3.2 if mail of content pane is not matched with respect to the certain_text
cpane=cpane-1
3.3 The panes for which the mail is matched are displayed.

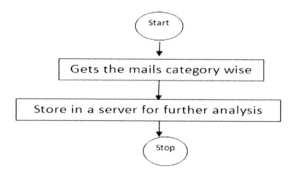

FIGURE 4.5 Statistics functionality module.

Figure 4.5 is a pictorial representation of activities involved in this module.

The demonstration of login, similarity of certain_Text against the mails with respect to panes, displaying mails based on panes in the statistics module is described in terms of pseudotypes and pictorial manner. The flow of these modules is mentioned below to achieve the intended theme (Figure 4.6).

The objective of the intended theme is described in terms of entities (rectangles) as modules and activities of each entity as use cases (ovals) (Figure 4.7).

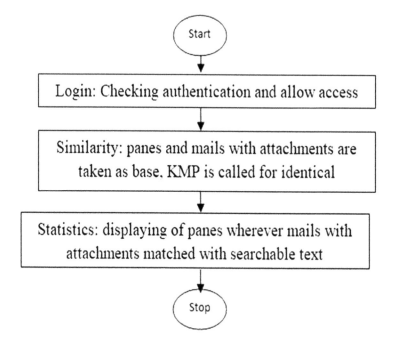

FIGURE 4.6 Proposed approach and its steps.

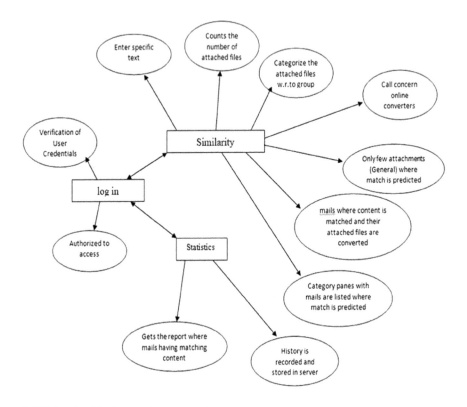

FIGURE 4.7 ER diagram for proposed system.

4.4 RESULTS

Here, each module is demonstrated in terms of output windows (Figure 4.8):

a. **Login:** The user has to be checked in and granted access to the resources of mail service provider. The below are snaps of it (Figures 4.9 and 4.10).

b. **Similarity Checking:** It gives certain_text as input, mails with attachments are counted, panes are also counted, mails whose attachments matched with the given text are tracked, and their panes are notified (Figures 4.11 and 4.12).

The options mentioned in Figure 4.13 are [Any time ▾], [Has attachment], and [To me]. In this window, only doc is the attachment file type is chosen.

If the type of attachment is either doc or pdf or both, the below window is enabled (Figure 4.14).

There are specific tools that transform a given file of one type to another type. That list is demonstrated below (Table 4.1):

c. **Statistics:** The panes and the mails that are matched with certain_text are displayed pane-wise (Figure 4.15).

Output of Login: Allow Access. type the pattern	→	Output of Similarity: Pattern inputted checked against Panes mail attachments	→	Output of Statistics: Displaying matched mails w.r.to panes

FIGURE 4.8 Sequence of module outputs in the intended approach.

FIGURE 4.9 Validation of user credentials.

FIGURE 4.10 Inbox.

FIGURE 4.11   Asking for input pattern to search.

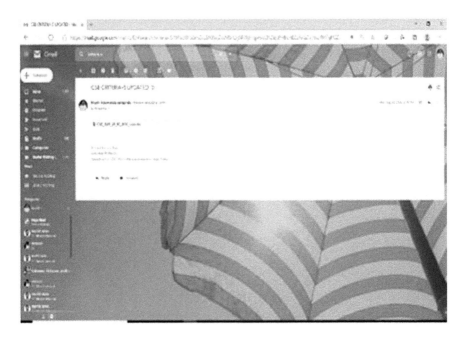

FIGURE 4.12   Mails that have matched with input pattern.

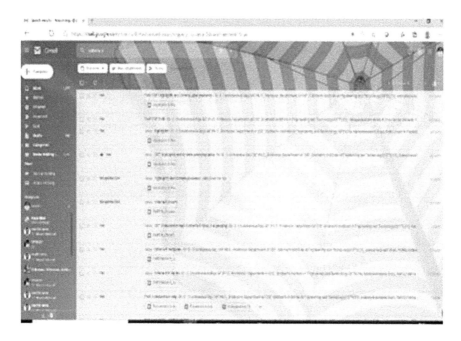

FIGURE 4.13 Output of proposed approach.

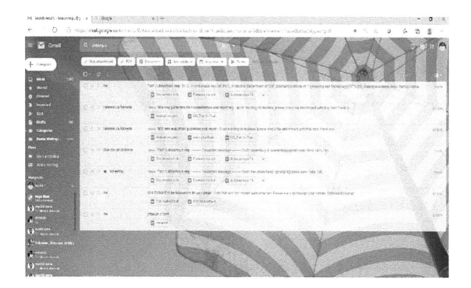

FIGURE 4.14 List of mails whose attachments matched with the input pattern.

TABLE 4.1
Specific Online Free Translators

Source	Destination	URL
jpeg	text	https://www.newocr.com/
Pdf	text	https://pdftotext.com/
Audio	text	https://www.kukarella.com/audio-to-text-converter
Video	text	https://vocalmatic.com/video-to-text-converter

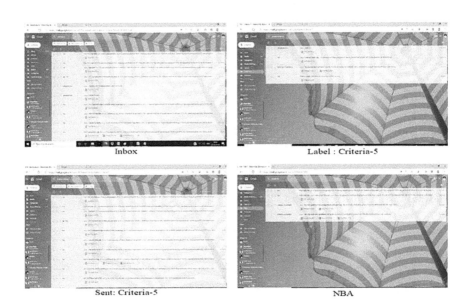

FIGURE 4.15   The proposed mail service in proposed output mode.

To understand how much user friendly between the conventional vs. intended approach is defined as below:

Here, the type of method and the speed level of the resulting output, respectively, are represented by X-axis and Y-axis.

The diagram in Figure 4.16 specifies two approaches where the first tower denotes the level or percentage of user-friendliness supported by the traditional mail system and the second tower mentions the extraordinary level of satisfaction by the defined approach.

The performance of this proposed approach over traditional is shown in the below graph (Figure 4.17).

We discussed both techniques when compared; the time consumed is more as well as performance is less, whereas the satisfactory time is consumed by the proposed approach and is expected to run more than five times the existing method. Here, based on panes available, the traditional approach consumes an increased amount of time whereas the proposed approach performs more processing and displays pane-wise mails. Hence, the proposed methodology is more efficient.

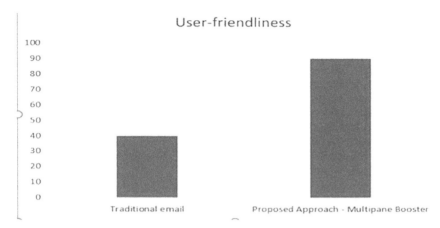

FIGURE 4.16 User-friendliness for traditional vs. proposed approach.

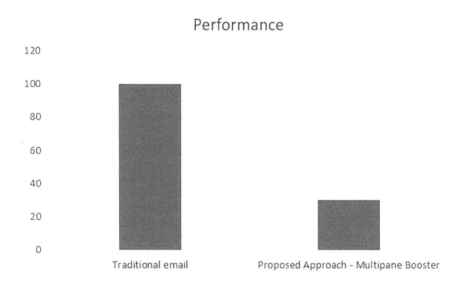

FIGURE 4.17 Performance for traditional vs. proposed approach.

4.5 CONCLUSION

This article demonstrates mails whose content is matched with specific text provided in usual mode and mails whose attachments are also matched with the given text. The mails are distributed according to the panes provided. The modules identified are login, similarity search, and statistics module. The output of one module is given as input to the other module in order to display the mails with respect to the content panes categorically. The details with respect to the factor performance and user friendliness are considered and proved that intended approach is far better than the traditional approach. In future, the input options may be more customized which may affect the output.

REFERENCES

[1] Kavin Tolly, Top 5 email security issues to address in 2019, December, 2019, https://searchsecurity.techtarget.com/tip/Top-5-email-security-issues-to-address-in-2019

[2] Staff Writer, 5 most common email security issues businesses face, December 2019, https://www.isoutsource.com/blog/5-common-email-security-issues-businesses-face

[3] Email Security Issues, 2021, https://www.mimecast.com/content/email-security-issues/

[4] Advantages and Disadvantages of Email, 2021, https://www.time-management-success.com/advantages-and-disadvantages-of-email.html

[5] Anton vdovin, The advantages and disadvantages of email for communications in a company, February 2020, https://www.alert-software.com/blog/the-advantages-and-disadvantages-of-email

[6] Networks and communications, email, GCSE, 2021, https://www.bbc.co.uk/bitesize/guides/zghfr82/revision/3

[7] Ann Lantz, Heavy users of electronic mail, November 2009, https://www.tandfonline.com/doi/abs/10.1207/s15327590ijhc1004_4

[8] Laura Garton & Barry Wellman, Social impacts of electronic mail in organizations: A review of the research literature, May 2016, https://doi.org/10.1080/23808985.1995.11678923

[9] Karen Renaud, Judith Ramsay & Mario Hair, "You've got e-mail!" … shall I deal with it now? Electronic mail from the recipient's perspective, June 2010, https://doi.org/10.1207/s15327590ijhc2103_3

[10] Find and replace text within an email message or item, 2019, https://support.microsoft.com/en-gb/office/find-and-replace-text-within-an-email-message-or-item-ee05a4ac-6c30-4687-8098-84f4390d2d6f

[11] Heinz Tschabitscher, How to search inside a message in outlook, February 2020, https://www.lifewire.com/search-inside-message-outlook-1173759

[12] Ricardo Vilhena, Search for string in the body of an e-mail, June 2019, https://support.google.com/mail/thread/5571070?hl=en

[13] A. Ahad, S. B. Yalavarthi, & M. A. Hussain (2018). Tweet data analysis using topical clustering. *Journal of Advanced Research in Dynamical and Control Systems*, 10(9 Special Issue), 632–636.

[14] A. Ajay Kumar, T. R. Kumar, & T. Bala Akhil Reddy (2018). Human resource management leave and tour management data retrieval system. *International Journal of Engineering and Technology (UAE)*, 7(2), 186–188.

[15] S. Anusha, D. Satish Kumar, P. Bindu, D. S. Rao, & H. Niranjan (2018). A descriptive study of customer's opinions on factors influencing purchasing decisions in corporate retail stores in a city of Vijayawada. *International Journal of Engineering and Technology (UAE)*, 7(4.10 Special Issue 10), 1029–1036.

[16] P. Balaji, D. Haritha, & O. Nagaraju (2018). Opinion mining based reputational trust assessment in e-commerce applications. *Journal of Advanced Research in Dynamical and Control Systems*, 10(2), 368–376.

[17] S. Bhavika, B. Prema Sindhuri, & G. Bhavana (2018). Spam detection using semantic web in mail services. *International Journal of Engineering and Technology (UAE)*, 7(2), 44–47. doi: 10.14419/ijet.v7i2.7.10255.

[18] D. B. Dasari, & K. V. G. Rao (2018). Context similarity strategy for text data plagiarism detection. *International Journal of Engineering and Technology (UAE)*, 7(2), 14–17. doi: 10.14419/ijet.v7i2.32.13517

[19] D. B. Dasari, & K. Venu Gopala Rao (2018). Semantic relevance scale for text data plagiarism detection. *Journal of Advanced Research in Dynamical and Control Systems*, 10(1 Special Issue), 811–819.

[20] Daulet Nurmanbetov, Cutting edge semantic search and sentence similarity, May 2020, https://towardsdatascience.com/cutting-edge-semantic-search-and-sentence-similarity-53380328c655

[21] Fangfang Liu, Lei Wang, Jie Yu, Context-aware similarity search of web service, 2012, https://ieeexplore.ieee.org/document/6221715

[22] G. Divya, & F. Noor Basha (2018). Implementation of DNA based cryptography using OTP random key generation process. *Journal of Advanced Research in Dynamical and Control Systems*, 10(2), 481–490.

[23] R. Dixit, & K. Ravindranath (2018). Encryption techniques & access control models for data security: A survey. *International Journal of Engineering and Technology (UAE)*, 7(1.5 Special Issue 5), 107–110.

[24] S. S. Gadde, R. K. S. Ganta, A. S. A. L. G. Gopala Gupta, K. Raghava Rao, & K. R. R. Mohan Rao (2018). Securing internet of things (IoT) using honeypots. *International Journal of Engineering and Technology (UAE)*, 7, 820–824.

[25] A. Gopi, & M. Kameswara Rao (2018). Survey of privacy and security issues in IoT. *International Journal of Engineering and Technology (UAE)*, 7, 293–296.

[26] S. Hrushikesava Raju, L.R. Burra, S.F. Waris, S. Kavitha, & S. Dorababu (2021). *Smart eye testing. Advances in Intelligent Systems and Computing, ISCDA 2020*, K L University, Guntur, Andhra Pradesh, 1312 AISC, 173–181, https://doi.org/10.1007/978-981-33-6176-8_19.

[27] Praveen Tumuluru, S. Hrushikesava Raju, C.H.M.H. Sai Baba, S. Dorababu, & B. Venkateswarlu, *ECO friendly mask guide for corona prevention, IOP Conference Series Materials Science and Engineering*, SRITW, Warangal, Telangana, India, 981, 2, 10.1088/1757-899X/981/2/022047

[28] C.H.M.H. Sai Baba, S. Hrushikesava Raju, M. V. B. T. Santhi, S. Dorababu, & Saiyed Faiayaz Waris, *International currency translator using IoT for shopping, IOP Conference Series Materials Science and Engineering*, SRITW, Warangal, Telangana, India, 981, 4, doi: 10.1088/1757-899X/981/4/042014.

[29] N Sunanda, S. Hrushikesava Raju, Saiyed Faiayaz Waris, & Ashok Koulagaji, *Smart instant charging of power banks smart instant charging of power banks, IOP Conference Series Materials Science and Engineering*, SRITW, Warangal, Telangana, India, 981, 210. doi: 1088/1757-899X/981/2/022066

[30] Radha Mothukuri, S. Hrushikesava Raju, S. Dorababu, & Saiyed Faiayaz Waris, *Smart catcher of weighted objects smart catcher of weighted objects, IOP Conference Series Materials Science and Engineering*, SRITW, Warangal, Telangana, India, 981, 2, doi: 10.1088/1757-899X/981/2/022002.

[31] Modepalli Kavitha, S. Hrushikesava Raju, Saiyed Faiayaz Waris, & Ashok Koulagaji, *Smart gas monitoring system for home and industries, IOP Conference Series Materials Science and Engineering*, SRITW, Warangal, Telangana, India, 981, 2, doi: 10.1088/1757-899X/981/2/022003.

[32] S. Hrushikesava Raju, Lakshmi Ramani Burra, Saiyed Faiayaz Waris, & S. Kavitha, *IoT as a health guide tool, IOP Conference Series, Materials Science and Engineering*, SRITW, Warangal, Telangana, India, 981, 4, doi: 10.1088/1757-899X/981/4/042015.

[33] S. Hrushikesava Raju, Lakshmi Ramani Burra, Ashok Koujalagi, & Saiyed Faiayaz Waris, *Tourism enhancer app: user-friendliness of a map with relevant features, IOP Conference Series, Materials Science and Engineering*, SRITW, Warangal, Telangana, India, 981, 2, doi:10.1088/1757-899X/981/2/022067.

[34] S. Hrushikesava Raju, M. Nagabhushana Rao, N. Sudheer, & P. Kavitharani, Agri-Iot: A sustainable environment for improvement of crops revenue in the field of agriculture using IoT, *International Journal of Engineering & Technology (UAE)*, 7 (2.32), 439–442, ISSN: 2227-524X, doi: 10.14419/ijet.v7i2.32.15735.

[35] S. Hrushikesava Raju, M. Nagabhushana Rao, N. Sudheer, & P. Kavitharani, IOT based home automation system with cloud organizing, *International Journal of Engineering & Technology (UAE)*, ISSN: 2227-524X, 7 (2.32), 412–415, doi: 10.14419/ijet. v7i2.32.15728.

[36] S. Hrushikesava Raju, M. Nagabhushana Rao, N. Sudheer, & P. Kavitharani, Quick identification of specific activity by processing of large-size videos using advanced spotter, *International Journal of Engineering & Technology (UAE)*, 7 (2.32), 355–358 ISSN: 2227-524X, doi: 10.14419/ijet.v7i2.32.15712.

[37] S. Hrushikesava Raju, M. Nagabhushana Rao, N. Sudheer, & P. Kavitharani, Visual safe road travel app over google maps about the traffic and external conditions, *International Journal of Engineering & Technology (UAE)*, 7 (2.32), 283–285, ISSN: 2227-524X, doi: 10.14419/ijet.v7i2.32.15697.

[38] Raju S. Hrushikesava, & M. Nagabhushana Rao, "Pattern matching using data preprocessing with the help of one time look indexing method", *International Journal of Pharmacy and Technology*, 8(3), 18395–18407, 2016, ISSN: 0975-766X.

5 Eyesight Test through Remote Virtual Doctor Using IoT

S. Hrushikesava Raju
Koneru Lakshmaiah Education Foundation, Guntur, India

Lakshmi Ramani Burra
PVP Siddhartha Institute of Technology, Vijayawada, India

Saiyed Faiayaz Waris
Vignan's Foundation for Science Technology and Research, Guntur, India

V. Lakshmi Lalitha, S. Dorababu, and S. Kavitha
Koneru Lakshmaiah Education Foundation, Guntur, India

CONTENTS

5.1 INTRODUCTION

The motto of making the novel approach that tests the human eye's vision is to automate the entire process. To cater to this kind of service, the Internet of Things technology uses a customized vision sensor loaded with normal eyesight parameters and a communication sensor that directs the sight values to the app designed, directing the spectacle store where the spectacle is to be purchased in the online mode. Once spectacle is selected, the customer may get feedback of that spectacle through the opinions from the friends (using social media). Hence, for shopping the spectacle in the online mode and getting the nearest shop's confirmation with delivery details, all activities are made automatic. In the earlier traditional approaches and semi-structured cases, the human effort as well as involvement is more than the proposed approach.

DOI: 10.1201/9781003156123-5

The novel approach has significant advantages, such as reducing time in terms of waiting for doctor's check-up, waiting at the spectacle shop, and money in terms of additional charges such as transport and miscellaneous activities. The traditional is the purely manual approach where certain time is taken to check the eyes, wait for the turn before taking check-up, and take care personally during these Covid-19 times. During the Covid second wave impact, the crowd at the shop, hospitals may lead to affect the consequences of ill health and lead to corona symptoms. To avoid such issues, automatic eye checks up would be used to provide the check-up efficiently and smoothly as expected.

The theme of the novel eye tester depends on the instructions or activities framed in the app. The app is used to automate the two significant activities such as eyesight checker and purchasing the spectacle in the online mode. The interaction is to be done one after the other in the row. This means the second module is automatically reverted after the first module processed and the report is obtained successfully.

The steps proposed in the novel approach are as follows:

1. Novel eye app is designed to ask users to register by asking their details and communication details. Once registered, you may get user and password details to get the service of the app, such as finding online shopping retail stores.
2. Two modules are defined: (1) an online eye check-up that uses predefined remote sensors and (2) a spectacle online purchase store where spectacle is to be purchased based on the models.
3. The two modules' pseudo procedures define the theme of the novel approach.
4. The comparison of proposed against the traditional, semi-structured approaches, based on time and human efforts or interventions, is also demonstrated.

5.2   RELATED WORK

The drawbacks of existing studies are helpful to discover the new model where enhancements or novel structure are to be discussed. The information provided in the sources specified in [1] to [8] is explanations and tips used as a guide for taking eyesight treatment. The key description mentioned in [9] to [12] illustrates many diseases for the eye and gives ways to solve such problems. Data extraction from the sources mentioned in [13] to [18] demonstrates certain methodologies and techniques for getting eyesight to be done somehow but not intended to automate the process. With regard to the information given in [19, 21–23] are describing different scenarios using IoT technology in the traditional way but they were not aligned towards achieving the full automation. The applications include social problems that are to be eliminated using IoT. As per the data given in [20], the concern is providing security and making storage confidentiality. In the view of the information provided in [24–26], the information provided is in the cloud in terms of storage as well as security using RSA, and multifactor authorization. With regard to the article as in [27], the consequences of the pervasive environment are described and making awareness of hybrid context studies and their framework is depicted. As per demonstration in [28], the fire environment is handled in a smart manner using IoT and its drawbacks, side effects are also listed under study. This would be a quick responsive approach in the emergency environment. In [29], the agriculture field is controlled and is monitored using a novel indexing approach for decision-making.

Regarding information given in [30], the gathering of temperature using specific type tube arrays and their impact is noticed in this study. Regarding sources mentioned in [31–35], the descriptions are given using sensors, communication among them, generation of reports for further decision or action to take up, reducing time complexity, and automating the process as much as possible. Regarding the source mentioned in [36], the approach uses an indoor environment as a point cloud, and PCOC provides the sensor device connectivity as a LoS detection technique. The efficiency of this approach is verified by ray-voxel intersection technique. In the view of the source specified in [37], the various devices in terms of sensors, scanners, and other useful forms are built to capture customer's eye focus on the items in the retail store and produce the statistics from which Fujitsu Forum increases sales of the items. In the source mentioned in [38], the sight laser detection model is designed that emits light from laser emitter through the Charge Couple Device (CCD) detector, assess the attenuation and turbulence, error rate are to be computed by various parameters involved in the double-pass propagation. Also, the MATLAB coding systems are used to detect the monostatic or bistatic type. The parameters used here are for the improvement of the budget link of the model. In the source mentioned in [39], the user's eye focuses on the line of sight on the eyeglasses and detects the angle based on four angles on the eyeglasses. The angle is detected using a CCD camera against the Field of View. In this, the error rate is observed as less than 1.5 degrees. The micro-fabricated optical sensors are used for LOS measurement and are helpful in extracting point of sight using specific selection of objects and their tuning functionality. According to the source from [40], the vision sensors use a camera as a primary unit that captures the image for the presence, orientation, and accuracy of that specific part. The two models, monochromatic and color models, are demonstrated to process the accuracy of the parts in the image. In the source defined in [41], the Diabetic Retinal (DR) and Sight-Threatening DR (STDR) are determined using the NM-based approach to determine the sensitivity and specificity parameters. The related PPV and NPV are also determined in assessing the DR and STDR. In the aspect of [42], the approach such as Diabetic Retinopathy (DR) is to be detected using the techniques such as nonmydriatic fundus photography, 7-field fundus photography, and smart phone fundus photography. These techniques are helpful in evaluating the specific parameters which are of interest. In the view of the source specified in [43], the AI is used as an automated software that determines the parameters such as sensitivity, specificity, and other related terms for models such as diabetic retinopathy and STDR. Here, the smart phone-based system is used and is helpful in ophthalmologist's grading.

Regarding the source specified in [44], the impaired vision of the adults is because of direct diabetic retinopathy. The smartphone-based camera is preferred for digital retina images. This device would help determine the factors such as affordable, accessibility, and easier to use than other traditional approaches used.

Regarding the source described in [45], impaired vision is focused on the global population due to diabetic retinopathy and requires life-long monitoring to predict early. The different techniques proposed, such as Smartphone Photography and other photography, are applied to compute DR and STDR. The remote interpretation and timely referral of such objects are sent to ophthalmologists. The advances in referrals support multi-modal and ultra-wide imaging create new avenues for assessing the DR. The various fundus photography are used to screen the DR and manage it in the world. In the source defined in [46], the imaging devices and tools in the ophthalmologists are used to examine the close interaction between the eye and tester. The high quality of the eye is managed via many parameters. Many newer tools are devised time being in this field to examine the eyes. In view of the source demonstrated in [47], smartphone retinal photography is proposed in screening the retina over the eye to determine diabetic retinopathy and other related issues. The primary setting called usability testing (retina scope) is required to improve quality service using medical assistants. They refine iteratively in the rounds based on results. Regarding the source specified in [48], diabetic retinopathy causes severe loss of eyesight in the patients. To prevent the loss early, NPDR is proposed. The advancements in DR would include with and without mydriasis, imaging in multiple filed or single fields, and the use of ultra wide-field screening or other field screening. In this, many such technological advancements, evidence gaps, and studies enhance the DR screening. With respect to the source specified in [49], the study is in the northeast of Italy area called Padova where DR is screened using telemedicine screening. The report produce based on DR incidence, Maculopathy, rate of progression to SDTR, and optimal screening interval with no DR in the patients with the first examination.

Regarding the source elaborated in [50], the review was done over 50 countries in Asia about diabetic retinopathy which is a severe cause of the eye is impaired. The screening of DR is a cost-effective approach to reduce the severity of loss of sight. The review finally compares the affected population % and its country. After analysis, the effect rate is less compared with other country ICO guidelines. Regarding the source defined in [51], the full-field flicker ERG with RETeval device is used to assess the DR with subjects such as mydriasis free and a tool such as full-field electroretinogram. The ROC is visualized for the type of DR such as mild, moderate, and severe. Regarding the source demonstrated in [52], telemedicine is applied over databases such as PubMed, EMBASE, and other types to determine the levels of DR and DME.

The QUARAS-2 technique judges the quality of the study. The parameters such as sensitivity and specificity are pooled based on random effects. The group of ROCs is assessed to show the overall performance of the test. The heterogeneity is judged using meta-analysis and subgroup approaches. In the description of the source mentioned in [53], the combination of few measures such as retina structure as well as function that intervene diabetic retinopathy is better than modality state the two approaches such as (ETDRS-DR) severity $\geq$ level 53 (ETDRS-DR+), and RETeval DR Score >23.5 (RETeval+). Many works also concentrated on diabetic issues in the patients and suggested screening to avoid losing eyesight in the early stage.

The mentioned study or work demonstrates the usage of certain functionality or methodology to be adapted in shaping the Internet of Things in the actual cause for real-time processing of the applications.

5.3 PROPOSED WORK

In this, ER Diagram is depicted for eyesight testing using IoT, which also resembles the architecture of a novel approach. The modules specified are eyesight check-up using sensors and online selection and purchasing of spectacles.

The app designed and its flow of screens are demonstrated in Section 5.3 (Figure 5.1).

There were 3 modules identified in this architecture and are demonstrated as below:

A. **Eyesight Check-up (Virtual doctor and computer automation):** In this, the virtual doctor is created for each user. The left and right eye sights are checked by the remote computer connected to the app. That computer loaded with a remote lens generates a report of the sight of that client.

B. **Spectacles Online Shopping (Spectacle ordering):** Once the eyesight report is generated would propagate to the spectacle store and ask the user to fit the spectacle to the user's face, allowing for an opinion from the friends in the social media, and finalize the spectacle. The merchant located in the nearby location accepts the order and confirms the user's delivery date and time through the mobile and mail id.

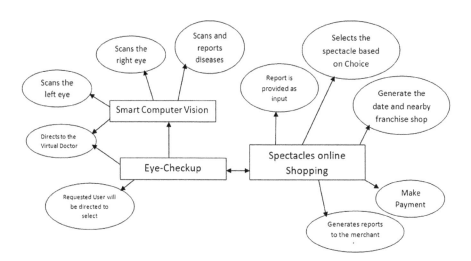

FIGURE 5.1   Novel eyesight approach - Eyesight check-up and online spectacle purchasing.

The algorithm for the **Eyesight Check-up module** is defined as below:

Step 1: Open the app, in which the authorized voice assistant directs the user
Step 2: Identify if any listed disease is predicted, provide a report
Step 3: Fix the eyes that focus on the refined remote camera behind the loaded vir-
         tual doctor which is activated with intended functionality
Step 4: Check left eye, and then the right eye
Step 5: Determine the distance violation from normal sight against the measured sight
Step 6: Generate the report

The algorithm for the **Spectacle Ordering module** is defined as below:

Step 1: Directed to select any one from the varieties of spectacles
Step 2: Once the model is fixed, ask the opinion from their social circle
Step 3: Once all is reviewed and self-analysis is made, finalize and make payment.
Step 4: Nearest spectacle shop merchant would accept and would deliver within the
         specified time frame.

   To visualize this flow, a flow chart is used and is depicted as follows (Figure 5.2):
   The results of each module are given as input to another module to process and
achieve the result. The results chapter focuses on the navigation of the screens in the
app designed. There the activities such as login, registration, remote eye virtual
doctor as an automated eye tester based on audio instructions in fixing the eyes over
the camera in the system, and directing to the next activity such as online spectacle
shopping where delivery details and merchant details are alerted once the frame is
selected by the user using social media.

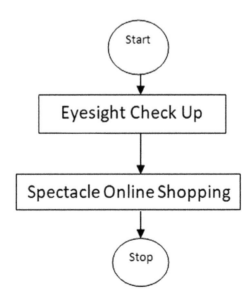

FIGURE 5.2   Novel approach interaction of modules.

5.4 RESULTS

The order of activities is specified in the interaction of the intended designed app.

The following are the screenshots of the app titles smart vision for eyesight (Figure 5.3):

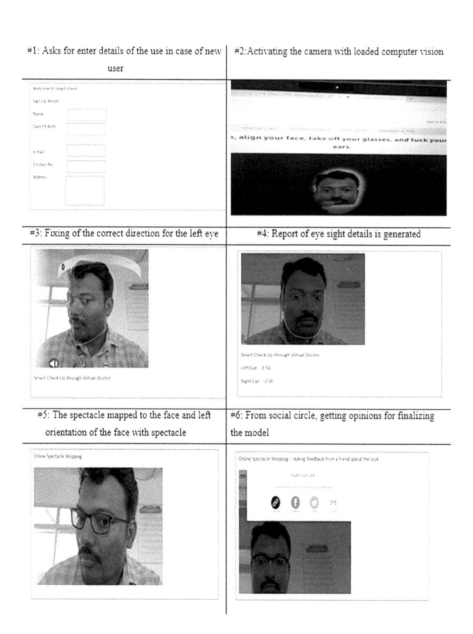

FIGURE 5.3 Order of integration of activities to automate the eyesight check-up and spectacle ordering.

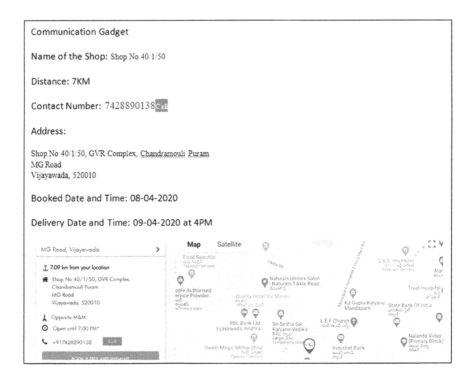

FIGURE 5.4 Distance of the shop and shop details, delivery details mentioned.

#7: Once getting the feedback, make a payment and report the nearest merchant spectacle store under that shopping franchise. The below is the snap of the delivery merchant detail; shipping details are demonstrated with mutual acceptance (Figure 5.4).

All the above activities of the novel app are done in a single place. It is based on an Internet connection. Nowadays, it became cheaper and is available with speed and reliability.

The number of users to be processed by the three kinds of approaches discussed in the article is demonstrated through the following graph.

From Figure 5.5, the key point is focused on the percentage of manpower needed for the mentioned approaches. The minimization of man power is almost very less in fully automated system of eyesight processing. Once the task starts, the entire process is automated till the report is generated and would get delivery details of the spectacle.

Figure 5.6 is demonstrated based on the number of users processed by the three approaches and the time consumed in processing the users in fully automated mode. Among the approaches provided, the fully automated approach considered here as the proposed approach would process many users more efficiently than the other two approaches that were time-consuming.

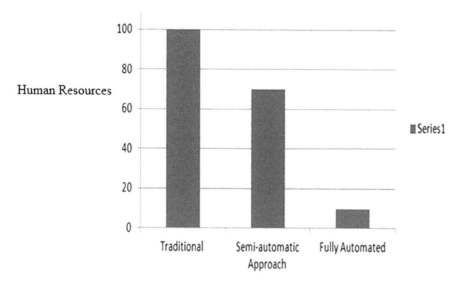

FIGURE 5.5 Number of human efforts in case of three approaches.

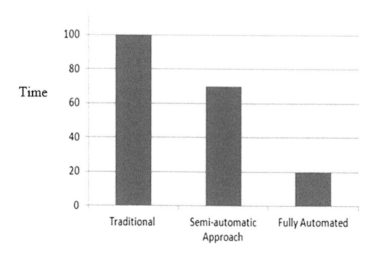

FIGURE 5.6 Performance in terms of automation in case of three approaches.

5.5 CONCLUSION

Here, the two modules play a key role which makes the system automatic. It reduces the human efforts in the system and maximizes the usage of resources in the online mode. The pseudo procedures of the modules demonstrate the working of the system. The two significant factors involved are the concentrated number of users and time. Both are significantly achieved as per the expectations of the novel app. The screens depicted focused on checking up the eyesight, generating the report on that, selecting spectacles from the models and getting delivery from the store merchants. During this pandemic and to avoid Covid-19 severity, this kind of automatic service-oriented apps are demanding and require other applications to be fully automated to reduce human efforts to save cost incurred in managing the system.

REFERENCES

[1] Understanding Vision, Take Part in the ZEISS Online Vision Screening Check and Test the Quality of Your Vision, ZEISS Online Vision Screening, Oct. 2017, https://www.zeiss.co.in/vision-care/better-vision/zeiss-online-vision-screening-check.html

[2] Eye Health, A Glossary of Eye Tests and Exams, June, 2020, https://www.webmd.com/eye-health/eye-tests-exams#1

[3] Michael X. Repka, Dan Gudgel, Home Eye Test for Children and Adults, March 2020, https://www.aao.org/eye-health/tips-prevention/home-eye-test-children-adults

[4] 6 Eye Tests in a Basic Eye Exam, 2011, https://allabouteyes.com/6-eye-tests-basic-eye-exam/

[5] David Turbert, Brenda Pagan-Duran, Eye Exam and Vision Testing Basics, Dec. 2018, https://www.aao.org/eye-health/tips-prevention/eye-exams-101

[6] Working Safely with Display Screen Equipment, https://www.hse.gov.uk/msd/dse/eye-tests.htm

[7] Liz Segre, Gary Heiting, What's an Eye Test?, https://www.allaboutvision.com/eye-test/free-eye-chart/

[8] Vision Screening, The College of Optometrists, Http Eye Charts and Visual Acuity Explained,https://www.allaboutvision.com/eye-test/free-eye-chart/s://guidance.college-optometrists.org/guidance-contents/knowledge-skills-and-performance-domain/examining-patients-who-work-with-display-screen-equipment-or/vision-screening/

[9] Michael Harkin, Ann Marie Griff, Visual Acuity Test, July 2012, https://www.healthline.com/health/visual-acuity-test

[10] Centers for Disease Control and Prevention, Basics of Vision and Eye Health, Common Eye Disorders and Diseases, https://www.cdc.gov/visionhealth/basics/ced/index.html

[11] Eye Health, Top Causes of Eye Problems, Reference, https://www.webmd.com/eye-health/common-eye-problems#1

[12] Richard E. Gans, The 5 Most Common Vision Problems and How to Prevent Them, Health Essentials, https://health.clevelandclinic.org/the-5-most-common-vision-problems-and-how-to-prevent-them/

[13] Robert Constantine, OTR/L, Vision Techniques for Eye Movement Disorders Associated with Autism, ADHD, Dyslexia & Other Neurological Disorders: Hands-On Assessments and Treatments for Children and Adolescents, RNV063660, https://www.pesi.com/store/detail/26169/vision-techniques-for-eye-movement-disorders-associated

[14] Daniel R. Gold, Eye Movement Disorders, in *Liu, Volpe, and Galetta's Neuro-Ophthalmology* (Third Edition), Diagnosis and Management, USA, 2019.

[15] R.J. Leigh, M. Gross, Eye Movement Disorders, in *Encyclopedia of Neuroscience*, 2009, https://www.sciencedirect.com/science/article/pii/B9780080450469010937

[16] Madhura A. Tamhankar, Eye Movement Disorders: Third, Fourth, and Sixth Nerve Palsies and Other Causes of Diplopia and Ocular Misalignment, in *Liu, Volpe, and Galetta's Neuro-Ophthalmology* (Third Edition), Diagnosis and Management, USA, 2019, https://www.sciencedirect.com/science/article/pii/B9780323340441000158, https://doi.org/10.1016/B978-0-323-34044-1.00015-8

[17] C.G. Chisari, A. Serra, Abnormal Eye Movements due to Disease of the Extraocular Muscles and Their Innervation, in *Reference Module in Neuroscience and Biobehavioral Psychology*, Diagnosis and Management, USA, 2017, https://doi.org/10.1016/B978-0-12-809324-5.01292-X

[18] Iliya V. Ivanov, Manfred Mackeben, Annika Vollmer, Peter Martus, Nhung X. Nguyen, Susanne Trauzettel-Klosinsk, Eye Movement Training and Suggested Gaze Strategies in Tunnel Vision - A Randomized and Controlled Pilot Study, https://journals.plos.org/plosone/article?id=10.1371/journal.pone.0157825, https://doi.org/10.1371/journal.pone.0157825

[19] M. Kavitha, Y. Manideep, M. Vamsi Krishna, P. Prabhuram, Speech Controlled Home Mechanization Framework Using Android Gadgets, *International Journal of Engineering and Technology (UAE)*, 7(1.1), 655–659.

[20] Sunanda Nalajala, et al., *Light Weight Secure Data Sharing Scheme for Mobile Cloud Computing*, Third International Conference on I-SMAC (IoT in Social, Mobile, Analytics and Cloud) (I-SMAC). IEEE, 2019.

[21] Modepalli Kavitha, Singaraju Srinivasulu, Kancharla Savitri, P. Sameera Afroze, P. Akhil Venkata Sai, S. Asrithl. (2019). Garbage Bin Monitoring and Management System Using GSM, *International Journal of Innovative and Exploring Engineering* 8(7), pp. 2632–2636.

[22] M. Kavitha, et al., *Wireless Sensor Enabled Breast Self-Examination Assistance to Detect Abnormality*, International Conference on Computer, Information and Telecommunication Systems (CITS). IEEE, Colmar, France, 2018.

[23] C.S. Kolli, V.V. Krishna Reddy, Kavitha, M. "A Critical Review on Internet of Things to Empower the Living Style of Physically Challenged People" *Advances in Intelligent Systems and Computing*, Springer, Singapore, IJRTE, 2020. 603–619.

[24] Sunanda Nalajala, Ch Pratyusha, A. Meghana, B. Phani Meghana, "Data security using multi prime RSA in cloud," *International Journal of Recent Technology and Engineering*, 7(6S4), 110–115, April 2019, ISSN: 2277-3878

[25] Sunanda Nalajala, et al., "*Data Security in Cloud Computing Using Three-Factor Authentication*," International Conference on Communication, Computing and Electronics Systems, Springer, Singapore, IJRTE, 2020.

[26] N. Sunanda, N. Sriyuktha, P.S. Sankar, "Revocable Identity Based Encryption for Secure Data Storage in Cloud," *International Journal of Innovative Technology and Exploring Engineering*, 8(7), 678–682.

[27] J. Madhusudanan, S. Geetha, V. Prasanna Venkatesan, U. Vignesh, and P. Iyappan, "Hybrid Aspect of Context-Aware Middleware for Pervasive Smart Environment: A Review," *Mobile Information Systems*, 2018, Article ID 6546501, 16 pages, 2018, 10.1155/2018/6546501.

[28] Lakshmana Phaneendra Maguluri, Tumma Srinivasarao, R. Ragupathy, Maganti Syamala, and N.J. Nalini, "Efficient Smart Emergency Response System for Fire Hazards using IoT," *International Journal of Advanced Computer Science and Applications (IJACSA)*, 9(1), 2018.

[29] M.S. Mekala and P. Viswanathan, "(t, n): Sensor Stipulation with THAM Index for Smart Agriculture Decision-Making IoT System," *Wireless Personal Communications*, 111, 1909–1940, April 2020. doi: 10.1007/s11277-019-06964-0.

[30] K.G. Kumar, B.S. Avinash, M. Rahimi-Gorji, and J. Majdoubi, "Photocatalytic Activity and Smartness of TiO_2 Nanotube Arrays for Room Temperature Acetone Sensing, *Sensors*, 21(5), 1826, February 2020, 10.1016/j.molliq.2019.112418.

[31] G. Subba Rao, S. Hrushikesava Raju, Lakshmi Ramani Burra, Venkata Naresh Mandhala, P. Seetha Rama Krishna, "*Smart Cyclones: Creating Artificial Cyclones with Specific Intensity in the Dearth Situations Using IoT, SMART DSC-2021*, 18–19 February, KL University, Guntur, Springer Series.

[32] S. Hrushikesava Raju, Lakshmi Ramani Burra, Saiyed Faiayaz Waris, and S. Kavitha, "*IoT as a Health Guide Tool*," *IOP Conference Series, Materials Science and Engineering*, 981, 4, SRITW, Warangal, Telangana, 10.1088/1757-899X/981/4/042015.

[33] S. Hrushikesava Raju, Lakshmi Ramani Burra, Ashok Koujalagi, and Saiyed Faiayaz Waris, "*Tourism Enhancer App: User-Friendliness of a Map with Relevant Features*," *IOP Conference Series, Materials Science and Engineering*, 981, 2, SRITW, Warangal, Telangana, 10.1088/1757-899X/981/2/022067.

[34] N. Sunanda, S. Hrushikesava Raju, Saiyed Faiayaz Waris, and Ashok Koulagaji, "*Smart Instant Charging of Power Banks*," *IOP Conference Series Materials Science and Engineering*, 981, SRITW, Warangal, Telangana, 210.1088/1757-899X/981/2/022066.

[35] S. Hrushikesava Raju and M. Nagabhushana Rao, "Improvement of Time Complexity on External Sorting Using Refined Approach and Data Preprocessing," *International Journal of Computer Sciences and Engineering*, 4(11), 82–86, 2016, E-ISSN: 2347-2693.

[36] Ravi Sharma, Venkataramana Badarla, and Vinay Sharma, "PCOC: A Fast Sensor-Device Line of Sight Detection Algorithm for Point Cloud Representations of Indoor Environments," *IEEE Communications Letters*, 24(6), 1258–1261, June 2020, 10.1109/LCOMM.2020.2981058.

[37] Daniel Klaus, "Line-of-Sight Detection Technology", Demonstration Draws Attention, May 2014, https://blog-archive.global.fujitsu.com/line-of-sight-detection-technology-demonstration-draws-attention/

[38] Christophe Lecocq, Gilles Deshors, Olga Lado-Bordowsky, and Jean-Louis Meyzonnette "*Sight Laser Detection Modeling*," *Proc. SPIE 5086, Laser Radar Technology and Applications VIII*, August 21, 2003, USA, https://doi.org/10.1117/12.486055.

[39] Masataka Ozawa, Kota Sampei, Carlos Cortes, Miho Ogawa, Akira Oikawa, and Norihisa Miki, "Wearable Line-of-Sight Detection System Using Micro-Fabricated Transparent Optical Sensors on Eyeglasses," *Sensors and Actuators A: Physical, Volume*, 205(1), 208–214, Jan. 2014, https://doi.org/10.1016/j.sna.2013.11.028.

[40] Detection based on "Camera Images" What Are Vision Sensors? https://www.keyence.co.in/ss/products/sensor/sensorbasics/vision/info/

[41] Vijayaraghavan Prathiba, Ramachandran Rajalakshmi, Subramaniam Arulmalar, Manoharan Usha, Radhakrishnan Subhashini, Clare E. Gilbert, Ranjit Mohan Anjana, Viswanathan Mohan, "Accuracy of the Smartphone-Based Nonmydriatic Retinal Camera in the Detection of Sight-Threatening Diabetic Retinopathy," *Indian J Ophthalmol*, 68(Suppl 1), S42–S46, Feb. 2020. doi: 10.4103/ijo.IJO_1937_19.

[42] M.E. Ryan, R. Rajalakshmi, V. Prathiba, R.M. Anjana, H. Ranjani, K.M. Narayan, T.W. Olsen, V. Mohan, L.A. Ward, M.J. Lynn, and A. M. Hendrick, "Comparison among Methods of Retinopathy Assessment (CAMRA) Study: Smartphone, Nonmydriatic, and Mydriatic Photography," *NCBI*, 122(10), 2038–2043, Oct. 2015. doi: 10.1016/j.ophtha.2015.06.011. Epub 2015 Jul 16.

[43] Ramachandran Rajalakshmi and Radhakrishnan Subashini, Ranjit Mohan Anjana, Viswanathan Mohan, "Automated Diabetic Retinopathy Detection in Smartphone-Based Fundus Photography Using Artificial Intelligence," *Eye*, 32(6), 1138–1144, Jun 2018. doi: 10.1038/s41433-018-0064-9. Epub 2018 Mar 9.

[44] Choon Han Tan, Bhone Myint Kyaw, Helen Smith, Colin S. Tan, and Lorainne Tudor Car, "Use of Smart phones to Detect Diabetic Retinopathy: Scoping Review and Meta-Analysis of Diagnostic Test Accuracy Studies," *J Med Internet Res*, 22(5), e16658, May 15, 2020. doi: 10.2196/16658.

[45] Ramachandran Rajalakshmi, Vijayaraghavan Prathiba, Subramanian Arulmalar, and Manoharan Usha, "Review of Retinal Cameras for Global Coverage of Diabetic Retinopathy Screening," *Eye*, 35(1), 162–172, Jan 2021. doi: 10.1038/s41433-020-01262-7.

[46] Amar Pujari, Gunjan Saluja, Divya Agarwal, Harathy Selvan, and Namrata Sharma, "Clinically Useful Smartphone Ophthalmic Imaging Techniques," *Graefes Arch Clin Exp Ophthalmol*, 259(2), 279–287, Feb 2021. doi: 10.1007/s00417-020-04917-z. Epub 2020 Sep 11.

[47] Patrick Li, Yannis M. Paulus, Jose R. Davila, John Gosbee, Todd Margolis, Daniel A. Fletcher, and Tyson N. Kim, "Usability Testing of a Smartphone-Based Retinal Camera among First-Time Users in the Primary Care Setting," *BMJ Innov*, 5(4), 120–126, Oct 2019. doi: 10.1136/bmjinnov-2018-000321. Epub 2019 Sep 19.

[48] Elizabeth Pearce and Sobha Sivaprasad, "A Review of Advancements and Evidence Gaps in Diabetic Retinopathy Screening Models," *Clin Ophthalmol*, 14, 3285–3296, Oct 14, 2020. doi: 10.2147/OPTH.S267521.

[49] Stela Vujosevic, Porzia Pucci, and Margherita Casciano, Anna Rita Daniele, Silvia Bini, Marianna Berton, Fabiano Cavarzeran, Angelo Avogaro, Annunziata Lapolla, and Edoardo Midena, "A Decade-Long Telemedicine Screening Program for Diabetic Retinopathy in the North-East of Italy, *J Diabetes Complications*, 31(8), 1348–1353, Aug. 2017. doi: 10.1016/j.jdiacomp.2017.04.010. Epub 2017 Apr 13.

[50] Louis Zizhao Wang, Carol Y Cheung, Robyn J Tapp, Haslina Hamzah, Gavin Tan, Daniel Ting, Ecosse Lamoureux, and Tien Yin Wong, "Availability and Variability in Guidelines on Diabetic Retinopathy Screening in Asian Countries," *Br J Ophthalmol*, 101 (10), 1352–1360, Oct. 2017. doi: 10.1136/bjophthalmol-2016-310002, Epub 2017 Mar 14.

[51] Yunkao Zeng, Dan Cao, Dawei Yang, Xuenan Zhuang, Honghua Yu, Yunyan Hu, Yan Zhang, Cheng Yang, Miao He, and Liang Zhang, "Screening for Diabetic Retinopathy in Diabetic Patients with a Mydriasis-Free, Full-Field Flicker Electroretinogram Recording Device," *OC Ophthalmol*, Jun. 2020; 140(3): 211–220. doi: 10.1007/s10633-019-09734-2, Epub 2019 Nov 12.

[52] Lili Shi, Huiqun Wu, Jiancheng Dong, Kui Jiang, Xiting Lu, and Jian Shi, "Telemedicine for detecting Diabetic Retinopathy: A Systematic Review and Meta-Analysis, *Br J Ophthalmol*, 99(6): 823–831, June 2015. doi: 10.1136/bjophthalmol-2014-305631, Epub 2015 Jan 6.

[53] Mitchell G Brigell, Bryce Chiang, Yauguang Maa, and C. Quentin Davis, "Enhancing Risk Assessment in Patients with Diabetic Retinopathy by Combining Measures of Retinal Function and Structure, *Transl Vis Sci Technol*, 9(9), 40, Aug 26, 2020. doi: 10.1167/tvst.9.9.40, eCollection 2020 Aug.

6 Recent Trends and Advances in Deep Learning-Based Internet of Things

Deepinder Kaur and Gaganpreet Kaur
Chandigarh University, Gharuan, India

Gurdip Kaur
University of New Brunswick, New Brunswick, Canada

CONTENTS

6.1   INTRODUCTION

The Internet of Things (IoT) is also referred to as the web of things [1]. It refers to all everyday objects or devices that can be connected to the Internet and have some kind of intelligence that allows data to be exchanged among various devices. This intelligence is mostly in the form of sensors that generate the data at various intervals of time. According to one survey, nearly 50 billion devices are predicted to be connected with IoT in 2020 only [2]. These devices may be connected through the public

DOI: 10.1201/9781003156123-6

or private network. For instance, smart wearable devices such as fitness bands, smart-watches, and smart shoes are becoming popular day by day.

Moreover, these IoT smart devices have proven to be helpful in critical situations like women's security. Women can wear tiny smart lockets or bands that sense the body parameters at regular intervals and send signals or messages to pre-registered contacts that may save the females in unfortunate incidents like rape or molestation [3]. Furthermore, IoT is beneficial in the healthcare industry, especially for disabled and old people. These smart devices send data to hospitals or their relatives continuously so that they can be saved in dangerous situations. Similarly, other areas such as smart home devices, smart agricultural devices, and many more are assisting human beings in living smart lives. The intelligent learning technique behind IoT can be any one out of regression, classification or clustering [4].

Out of various machine learning techniques, Deep Learning (DL) is a fantastic technique that works on collection of structured and unstructured data and helps in decision-making. Deep Learning has many application areas such as image analysis, pattern recognition, etc. Nevertheless, DL is proven to be the best technique in big data analytic solutions [5]. Currently, there is more generation of unstructured data in the form of audio signals or images for which traditional machine learning approaches are not enough. Moreover, Deep Learning uses artificial neural networks similar to human brains for bringing out the machine learning process. A variety of Deep Learning traditional and advanced models have been proven to be efficient in decision-making in a wide range of real-world applications.

Various data analytics techniques, for example, machine learning, can be used on this generated data to find novel information which may further help in decision-making. Traditional machine learning approaches used with IoT were only able to understand the structured data but not the audio signals and visual images [6]. However, Deep Learning is used in almost all IoT application areas [7]. Deep Learning has encouraged learning and analytics in various IoT domains [4]. Deep Learning uses a neural network that sometimes translates audio signals to a list of words used to find relevant information. Deep Learning or machine learning collaborated with IoT to overcome the security challenges [7]. In smart cities and smart campuses, a Deep Learning approach with IoT helps improve interaction and monitoring. Like In flood monitoring, Deep Learning helps IoT-enabled devices like smart cameras to capture drainage blockage images [8]. Deep Learning incorporated with IoT has saved millions of dollars in many areas. For example, in smart home devices, Microsoft and Liebherr applied Cortana DL for gathering the information from inside the refrigerator [9]. With Deep Learning, IoT devices may sense the food and medicine portion to be taken by a person. It may help decide whether the particular heartbeat or body temperature sensed by a smart gadget is critical or normal. Furthermore, DL is also used in the detection of cardiovascular diseases. IoT and DL combination in the area of well-being and healthcare have been proven to be God's gift for humans [10].

This chapter is organized into different sections. In Section 6.1, the terms IoT and Deep Learning are introduced along with their integrated applications. In Section 6.2, different advanced Deep Learning techniques have been explained. Section 6.3 gives insights about which Deep Learning techniques have been used in the different IoT applications, followed by the conclusion.

6.2   ADVANCED DEEP LEARNING TECHNIQUES

Deep Learning is a machine learning technique that works on traditional neural networks and is the most efficient. Moreover, it also develops various learning models using graph technologies with transformations between the neurons [11]. Deep Learning works on feature engineering better than conventional machine learning approaches to unsheathe the selective features with less man effort and domain knowledge [12] (Figure 6.1).

With the passage of time advanced Deep Learning techniques came into existence to enhance the scope and performance of applications and models. These advanced Deep Learning techniques are Unsupervised and Transmit learning, online learning, optimized techniques in DL, and Deep Learning in distributed systems. Moreover, DL frameworks incorporate modularized Deep Learning techniques, distribution and optimization techniques to uplift the system development and research [13].

6.2.1   UNSUPERVISED AND TRANSMIT LEARNING

The researchers and reviewers discussed unsupervised or self-taught learning ideas in different ways in the last few years' literatures [13–15]. In an unsupervised approach, a vast amount of data is trained to learn the features. For instance, training on a dataset of very large-scale images is provided to the deep sparse AE to learn the features of face detection using unsupervised techniques [14]. The transmit learning or transfer learning approach is done with pre-training the deep network such as CNN on large-scale datasets.

Different models are developed with an unsupervised learning approach and effectively outperform the other conventional learning approaches [16].

6.2.2   ONLINE LEARNING

In traditional Deep Neural Network (DNN) learning, the whole training of data is to be done in advance, but it is not possible in some cases due to the streaming of data from time to time. Thus, Online Deep Learning came into existence to avoid retail

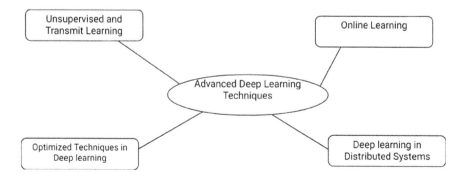

FIGURE 6.1   Deep Learning techniques.

and banking pipelines, including business value [13]. Online learning takes advantage of artificial intelligence with instinctive algorithms. In this technique, data is trained in sequence rather than batch processing in conventional approaches. Researchers have propped novel Hedge Back Propagation (HBP) for online updating the DNN parameters for online streaming scenarios. HBP automatically decides the time and approach to transform the network depth online [17].

6.2.3 OPTIMIZATION TECHNIQUES IN DEEP LEARNING

DNN is trained, which is known as the optimization process to observe the variables in the network that reduce the loss function [13]. Various optimization techniques used in Deep Learning are SGD (stochastic gradient descent) [18], Adagrad [19], Adadelta [20], and NAdam [21]. Well-designed random initialization of SGD with momentum can train DNN efficiently [18]. Adagrad is adaptive gradient-based learning that outperforms online learning in sparse gradient vectors [19]. Adadelta is another adaptive per-dimension learning method that modifies using first-order information over time and works efficiently for large datasets, powerful for noisy gradient information [20]. Similarly, NAdam is an amalgamation of Nesterov's accelerated gradient (NAG) and popular optimized algorithm Adam that improves the performance of the Deep Learning system to distribute the error signal throughout the model variables, hence resulting in better quality learned models [21].

6.2.4 DEEP LEARNING IN DISTRIBUTED SYSTEMS

Distributed Deep Learning has been implemented to expedite the training process through two concepts: data parallelism and model parallelism. Model parallelism is beneficial for larger Deep Learning models, while data parallelism is better for larger datasets [22]. Moreover, some researchers have combined both approaches to make a hybrid model that results in faster convergence [23].

6.3 APPLICATIONS OF IoT USING DEEP LEARNING

IoT is to convert the conventional manual things to intelligent ones using advanced technologies like sensors, Internet protocols, applications, or smart hardware gadgets [24]. To manage the enormous amount of data generated with IoT devices with applications, sensors, or microcontrollers, traditional machine learning approaches of data analytics are not sufficient. Thus, advanced approaches such as Deep Learning are proving to be blessings for such situations. Tiwari et al. [25] explained the main key factors which show that Deep Learning is an efficient approach for IoT applications. Yao et al. [26] discussed different challenges of using Deep Learning in IoT applications and the solutions of erecting effective and efficient IoT applications with the help of Deep Learning.

Currently, almost every area is covered by IoT with advanced machine learning approaches such as Deep Learning. Intelligent or smart devices work and generate data without human interference. Whether it is the medical field, farming,

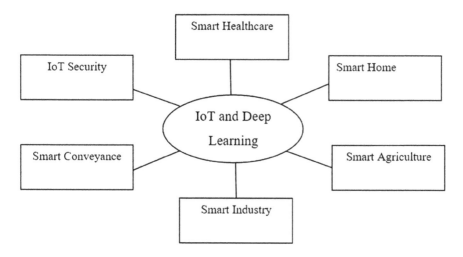

FIGURE 6.2   IoT with Deep Learning applications.

manufacturing industries, or home services, every region takes advantage of IoT and Deep Learning for more accurate, reliable, and secure services. For instance, in agriculture, these IoT applications help detect soil nutrients or water levels for different crops. Similarly, smart conveyance applications such as parking optimization, traffic monitoring, and motor navigation assist humanity in transportation. Besides this, there are ample IoT medical and security applications that are doing wonders through Deep Learning (Figure 6.2).

6.3.1   IoT Security

Due to their open access, IoT devices are more prone to malware attacks such as privacy theft, traffic consumption, and remote control. Moreover, IoT devices generate 3 V's, i.e., large volume, variety, and velocity data [27]. According to a VDC Research Group Inc. study, 60% of the challenges in developing connected devices are related to security requirements [28].

The mainly used techniques in this malware are encryption and obfuscation. Therefore, various machine learning malware detection methods were developed. However, these methods required lots of effort and took more time to detect the malware. Traditionally, machine learning-based methods were classified into static, dynamic, and hybrid detection [29] (Figure 6.3).

The process used for malware detection in IoT devices mainly consists of four steps, from building datasets to performing feature extraction. Then, machine learning models are trained, following which the final performance evaluation of trained models is done [30]. Therefore, Deep Learning-based malware detection methods were proposed to overcome this issue. Moreover, these Deep Learning methods for IoT security have shown efficient results with low resource consumption and unlimited input file size.

IOT Security

FIGURE 6.3 Applications of IoT security with Deep Learning [31, 32].

Huang et al. [33] proposed an Android-based malware detection system (R2-D2) with CNN by transforming android applications' classes.dex files into colorful images and, nevertheless, this developed model is too huge to be deployed on IoT devices. Al-Garadi et al. [34] explained the importance of IoT security with advanced machine learning and Deep Learning applications. Kaur et al. [35] surveyed many IoT software applications and hardware gadgets that collect data analytics using various factors and alarm the registered contacts in case of an unsafe situation. Zhongru Ren [29] proposed two Deep Learning-based Android malware detection techniques named as DexCNN and DexCRNN. These techniques were trained and assessed on the resampled classes.dex files of Android applications without any manual feature engineering. DexCNN method was inspired using TextCNN [28] and Malconv [36]. TextCNN model used CNN for performing text classification tasks while MAlconv performed windows malware detection with 94% accuracy [36]. Similarly, DexCRNN was developed by combining two kinds of neural networks, namely CNN and RNN. Nonetheless, unlike DexCNN initial embedding layer is not required in DexCRNN. These two Deep Learning methods were effective for Android IoT devices with no security expert intervention [37].

Abebe Abeshu Diro et al. [38] used the DL approach for detecting the distributed attack on the social Internet. They demonstrated that their deep model proved to be more efficient than the traditional machine learning approach. Aidin Ferdowsi et al. [39] developed a novel DL technique to detect cyber-attacks by dynamic watermarking of IoT signals. This concept of LSTM structure was utilized in this technique. This method allowed the IoT cloud center, which collects signals from IoT devices, to effectively authenticate the reliability of the signals, assisting in the prevention of various cyber-attacks for intensifying the security and durability of IoT. The authors proposed a framework that integrates SDN, IoT, and DL to develop an Intrusion Detection System (IDS). The detection component utilized the Restricted Boltzmann Machines (RBM), which outperformed traditional machine learning approaches.

Zhihong Tian et al. [40] proposed a web attack detection system that analyzes URLs, detects web attacks, and demonstrates more effective results than conventional machine learning approaches.

Jafari et al. [41] presented a wireless device recognition platform for enhancing IoT security using Deep Learning approaches by capturing Radio Frequency traces covering a broad range of SNR (Signal-to-Noise Ratio) levels. Roopak et al. [42] used current CICIDS2017 datasets and proposed Deep Learning models for detection of DDoS (Distributed Denial of Service) attacks and showed the accuracy of approximate 97% in contrast to other machine learning algorithms. Sagduyu et al. [43] presented novel approaches to protect the IoT networks using Deep Learning technique based on adversarial machine learning. Zahangir et al. [44] used unsupervised Deep Learning approaches, specifically AE and RBM, for network intrusion detection. Besides AE and RBM, they also used iterative K-means clustering and Unsupervised Extreme Learning Machine (UELM) for implementation. Abeshu and Chilamkurti [45] explained the different challenges for cybersecurity due to the increase in smart devices. They also proposed a new distributed Deep Learning approach of cyber-attack identification with fog-to-things computing. This technique outperformed the other conventional machine learning approaches in terms of scalability and false alarm rate.

6.3.2 SMART HEALTHCARE

Among the various applications of IoT, smart healthcare is proven as a blessing, especially for older persons. These healthcare applications include health monitoring and disease analysis applications. Undoubtedly, these applications and IoT devices produce a tremendous amount of data, and to handle this much amount of data, Deep Learning helps to extract the minute and hidden features and convert the big medical data to enhance human health [46]. There are ample IoT applications and devices developed to monitor health, which continuously produce data based on body parameters such as pulse rate, heartbeat, or body temperature. If the values are higher than threshold values, then help is provided to such persons. For this continuous data generation, Deep Learning models are implemented to extract important and useful data. One example of such IoT application is human activity recognition (HAR) [46], which analyzes health conditions and human activities.

Hammerla et al. [47] developed advanced Deep Learning techniques for HAR using CNN and LSTM to analyze the movement data of the person and then use these results to predict Parkinson's disease (Figure 6.4).

Another application is human action monitoring [48] based on Deep Learning architecture R3D (recurrent 3D (CNN) Convolutional Neural Network) and LSTM that withdraw temporal features for human action recognition. The other category is disease analysis applications like knee cartilage segmentation [49], in which the CNN approach is used to represent knee MRI scans in a hierarchal structure to encourage classification. An additional IoT application is SPHA (smart personal health advisor) that also uses CNN for psychological and physiological health monitoring and guidance [50]. Some other healthcare applications that use Deep Learning

FIGURE 6.4   Smart healthcare with Deep Learning [51, 52].

approaches for the extraction of important features are Pill image recognition [53] to recognize unconstrained pill images, skin lesion classification [54] to identify whether the skin lesion image is friendly or malevolent, medicine recognition [55] in which ST-Med-box can help the chronic patients to take correctly numerous medications.

Shreshth Tuli, Nipam Basumatary et al. [56] developed an application, "HealthFog" to diagnose heart diseases using ensemble DL and Edge computing. In this particular application, the model's training is done using heart patient data. Then the trained model can be used for the prediction of the output of real-time data input, i.e., whether the patient has heart disease or not.

Azimi et al. [57] developed a hierarchical computing architecture (HiCH) in which the system availability was enhanced efficiently with the help of local decision-making. HiCH was based on CNN, which ensures accuracy and less response time. Ali et al. [58] presented a smart healthcare monitoring system with DL which increased the diagnosis accuracy of heart diseases as compared to the conventional machine learning approaches. Alhussein et al. [59] proposed an automatic voice pathology monitoring system using two publicly available datasets that take voice samples from cell phone and perform the experiment. This system proved to be very effective for aged people. Sundaravadivel et al. [60] explained the importance of essential nutrients for the body. They proposed an automated nutrition monitoring system based on Deep Learning, namely Smart-Log. This mobile application uses Wi-Fi sensors for measuring food nutrition.

Muhammad et al. [61] described a detailed contemplation of DL methods for Brain Tumor Classification (BTC) with every method's strength and limitation. In addition to past literature, they have also mentioned future research scope in the same direction. Khan et al. [62] proposed an automated Deep Learning approach, "StomachNet," to stratify gastrointestinal infections. Nevertheless, they have also taken the help of two other metaheuristic algorithms: enhanced Crow Search and

Differential Evolution. However, the result accuracy is 99.46% that is better than other neural network paradigms. Gumaei et al. [63] proposed a multi-sensors-based substructure for recognition of human activity with the help of a hybrid DL model that uses the combination of SRU (Simple Recurrent Unit) and GRU (Gated Recurrent Unit) neural network architecture. This proposed approach can be considered beneficial for elderly and physically disabled persons based on multimodal body sensing factors. Cılasun et al. [64] implemented a DL technique for epilepsy seizure determination based on Deep CNN architecture with EEG (electroencephalogram) signals. Andres Oriz et al. [65] combined Deep Learning architectures for premature detection of Alzheimer disease. This research paper explored the classification construction methods that pertained to brain regions. Amin et al. [66] proposed a framework for healthcare that utilizes smart sensors and DL for decision-making smart city patients in which state of the patient is resolved by monitoring sensors and collecting body factor information such as gestures and EEG signals. This information helps in classification of normal and pathological patient.

6.3.3 SMART HOME

Domotics or smart home automation is yet another important IoT application that helps provide comfort and alleviation to human beings. The term smart homes mean that ubiquitous devices with the sensors used in home appliances can be broadly categorized into two divisions such as indoor localization and home robotics [67]. Indoor localization itself is a combination of many sub-applications, including elder monitoring, kids monitoring, and wireless intruder detection. However, there are various challenges in these applications, such as delay distortion and fading. At this point, Deep Learning helped in the development of these IoT applications along with the removal of the above challenges [68].

Many algorithms were proposed and showed efficient results. For instance, an algorithm SDELM (Semi-supervised Deep Extreme Learning Machine) given by Gu et al. [69] combines the benefits of Deep Learning, semi-supervised learning, and extreme learning machine, and thereby gives effective results on the localization and minimizes the calibration effect. Similarly, Mohammadi et al. [70] proposed another semi-supervised approach, DRL (Deep Reinforcement Learning) that uses variations auto-encoders as the inference engine and shows optimal policies. The second division is home robotics in which home gadgets connect to the Internet and provide intuitive services to humans. Home Robotics consists of different applications like object recognition, home navigation, human–robot interaction, etc. (Figure 6.5)

Different Deep Learning architectures such as autonomous navigation have been developed in which CNN is implemented into IoT robot devices for object recognition and markers using RGB-depth cameras [71]. Furthermore, Levine et al. proposed hand–eye coordination for robotics-gasping, which is also trained using CNN that foresee the probability of grasping using monocular camera images [72]. Another division is miscellaneous applications. Deep Learning techniques are popularly used in digital personal assistants such as Apple's Siri, Google Assistant, etc., which helps collect valuable data [73]. Faisal Mehmood et al. [74] used an object detection

FIGURE 6.5   Smart home applications with Deep Learning [75].

algorithm based on model view controller (MVC) architecture to create an intelligent home automation system. Valentina Bianchi et al. [76] designed a wearable sensor coupled with CNN that includes an Inertial Measurement Unit (IMU) and a Wi-Fi and section to transfer data for daily activity monitoring of the user high accuracy.

6.3.4 SMART CONVEYANCE

With the advancement in the Internet, IoT has also broadened its horizon in the field of transportation. Without any doubt, Deep Learning helped IoT applications in this area also. IoT and Deep Learning have been applied in different categories such as autonomous driving, traffic monitoring, and traffic prediction [67].

Intelligent or smart transportation is an integral part of any smart city. It includes basic vehicle management systems like motor navigation, automatic number plate recognition to parking optimization, and traffic signal control systems, accident prevention/detection [77] (Figure 6.6).

Sensors are being embedded into the vehicles for route optimization using applications such as Google maps. However, an enormous amount of data is generated during intelligent vehicle monitoring or traffic optimization, and to handle this, Deep Learning approaches have been proposed and show effective results. One such application developed was traffic flow prediction [78], in which a Deep Learning DBN model was used to seize the pictures of street traffic networks in each part. Similarly, LSTM-based traffic prediction model was proposed, which gave efficient results for forecasting the traffic that may help passengers decide the best route, mode, or time [79]. A LSTM model, RNN approach of Deep Learning was used. Another category for smart transportation is traffic monitoring, in which automatic monitoring of traffic can avoid accidents by sending warning messages to the drivers. Deep Learning algorithms used for traffic monitoring are used for object detection, face recognition,

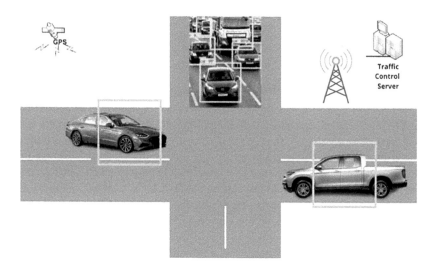

FIGURE 6.6   Smart conveyance applications with Deep Learning.

and object traffic [67]. Object detection can be applied for commuters' detection or on-road vehicle detection. Moreover, Deep Learning CNN and multi-scale strategy have been used in various models developed such as RPN (region proposal network) [80], YOLO (You Only Look Once) [81] for license plate recognition and SSD (Single Shot multibox Detector) [82] to enhance the speed and accuracy. Face recognition or object tracking has been mainly used in surveillance systems to track suspected people for safety monitoring. Similarly, Jiateng Yin et al. [83] developed an automated diagnosis network of VOBE (Vehicle On-board Equipments) for fast trains using the DL approach. They developed a DBN (Deep Belief Network) based on RBM. Finally, they concluded that the accuracy of fault diagnosis using DBN is more than the traditional approaches such as k-nearest neighbor (KNN) and ANN-BP (artificial neural network with back propagations). Aidin Ferdowsi et al. [84] discussed that Deep Learning solutions for intelligent transportation systems are more efficient than conventional cloud computing solutions in terms of reliability and security of the transportation environment. Zhu et al. [85] proposed joint quadruple decorrelation directional deep networks (JQD 3 Ns) for reidentification of pedestrian or vehicle images captured from different perspectives, which outperformed other traditional object identification algorithms. Ashraf et al. [86] presented an IDS based on DL for smart transportation systems to investigate suspicious network activities in autonomous vehicles based on LSTM autoencoder algorithms. This system attained 99% attack detecting accuracy on the car hacking dataset. Zhang et al. [87] developed a DL approach for determining the employment status of public transportation users using Smart Card Data, which may solve the problem of estimating travel demand. This approach used CNN architecture and categorized the passengers into six types according to employment status. Zhou. et al. [88] proposed the RMCS (Robust Mobile Crowd Sensing) framework for sensing the traffic load and latency with the help of DL and Edge computing. Pandey et al. [89] provided a detailed analysis of intelligent transportation systems' latest object detection techniques to improve vehicle automation and autonomous driving.

6.3.5 Smart Industry

IoT has spread its wings in all fields and the smart industry has also become an integral part of smart cities. IoT helps collect various parameters from industry such as humidity, temperature, and ph for online monitoring solutions [90]. The data is collected using various sensors that may help in controlling environmental pollution levels. Without any doubt, Deep Learning helped IoT in this field too. Manufacturing inspection and fault assessment are two broad categories in which Deep Learning provides help to IoT in the intelligent industry [67]. A DL classification model, such as a robust inspection system, was proposed using CNN to enhance the computing efficiency and deal with big data [91]. Another similar Deep Learning CNN-based model developed was surface integration inspection in which patch features were extracted and defect areas were predicted in less time and cost [92]. Fault diagnosis and identification and transformer fault diagnosis are two main Deep Learning fault assessment models in which wavelet-based CNN and CSAE (Continuous Sparse Auto Encoder) approaches were used to predict the faults [93, 94].

Jay Lee et al. [95] proposed a self-awareness and self-maintained machine system that can self-access its own degradation that may further help make smart maintenance decisions with the help of IoT and advanced CPS (Cyber Physical Systems). Olivier Janssens et al. [96] came up with a feature learning model for condition (Figure 6.7).

Monitoring gleaned from CNN (Convolutional Neural Networks). This approach surpassed the accuracy results obtained with the traditional approach, i.e., feature engineering-based approach in fault detection of rotating machines. Ren et al. [97] experimented with defect segmentation on surfaces with a 0% error escape rate using a generic-based Deep Learning approach.

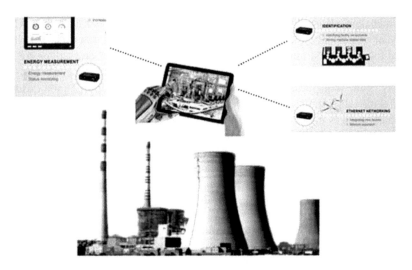

FIGURE 6.7   Smart industry applications with Deep Learning.

6.3.6 SMART AGRICULTURE

There are many application areas in which Deep Learning with IoT served best and made humans more comfortable [98]. The agriculture field has also acquired benefits from IoT with Deep Learning. Deep Learning can detect water and nutrient levels of the soil for various crops [99]. RNN has been utilized in various smart farming areas like crop yield estimation, LCC (land cover classification), weather prediction, phenotype recognition, and many more [100]. Similarly, CNN is used in plant disease detection due to its proficiency in image processing. Plant disease detection can be done using leaf image classification and pattern recognition [101]. The results with Deep Learning techniques are more efficient as compared to manual detection techniques.

CNN has also been used in weed detection using approaches such as K-means feature learning [102] and plant classification using ALexNet [103]. Likewise, GAN (Generative Adversarial Networks) is a novel type of neural network used in image processing and video generation [100, 104–108]. However, GAN is not widely being used in smart agriculture applications. Barth et al. [109] developed a technique used for the optimization of realism in synthetic images. Smart agriculture systems are mainly used to ameliorate harvesting (Figure 6.8).

Arrhul et al. [110] utilized the LSTM networks algorithm for soil monitoring to help the user grow the best suited or profitable crop on his/her land. Xue-Bo Jin et al. [111] developed a hybrid Deep Learning predictor that takes the help of an empirical mode decomposition (EMD) and divides the climate data into fixed component groups with different frequency characteristics to predict the various factors such as wind speed, temperature, and humidity, the data is trained with GRU (Gated Recurrent Unit) to achieve the prediction result.

Roopaei et al. [112] deployed a Cloud of Things-based automated irrigation system that collects data analytics like weather information, soil kind, and merges this information with data encapsulated with sensors estimating water resources level,

FIGURE 6.8   Smart agriculture applications with Deep Learning [113].

moisture, etc., that allow the farmers to utilize water and pesticides levels to maximize the crop production. Lee et al. [114] developed Deep Learning entrusted to fog nodes (DLEFN) algorithm that claimed to be an effective algorithm for FC-based smart agriculture environments for better service without delays.

Laha Ale et al. [115] proposed DenseNet (Densely Connected Convolutional Networks) and lightweight DNN to detect plant diseases more accurately with less computational resources (Table 6.1).

TABLE 6.1
IoT Applications and respective Deep Learning Technique Used

IoT Application Area	Proposed Model	Deep Learning Approach Used
IoT security	R2-D2 [33]	CNN
	DexCNN [29]	CNN
	DexCRNN [29]	CNN + RNN
	Dynamic Watermarking [39]	LSTM
	Intrusion detection system [116]	SDN based on RBM
	Wireless Device Identification [41]	CNN + DNN +RNN
	Network Intrusion Detection [44]	AE + RBM
Smart healthcare	HAR [47]	CNN and LSTM
	Human action monitoring [48]	R3D
	Knee cartilage segmentation [49]	CNN
	Smart personal health advisor [50]	CNN
	Pill Recognition System [53]	CNN
	Skin lesion detection [54]	CNN
	Health-Fog [56]	Edge computing+ DNN
	HiCH [57]	CNN
	Smart-Log [60]	Perceptron Neural Network
	StomachNet [61]	DCNN
	Hybrid Deep Learning model [63]	SRU-GRU neural network
	Epilepsy seizure detection [64]	DCNN
Smart home	(Semi-supervised Deep Extreme Learning Machine) SDELM [69]	DNN
	Indoor localization [70]	DRL
	Autonomous navigation [71]	CNN
	Hand–eye coordination for robotic gasping [72]	CNN
	Smart home automation system [74]	SSD
	Wearable device for daily monitoring [76]	CNN
Smart conveyance	Traffic flow prediction [67]	DBN
	Short-term Traffic prediction [77]	LSTM
	License Plate Recognition [81]	YOLO
	VOBE [83]	DBN+RBM
	joint quadruple decorrelation directional Deep networks JQD^{3}NS [85]	CNN
	Intrusion detection system [86]	LSTM
	Employment Status Approach [87]	CNN
	RMCS [88]	Edge computing
Smart industry	Robust Inspection System [91]	CNN
	Surface integration inspection [92]	CNN
	Fault diagnosis and identification [93]	Wavelet-based CNN
	Fault diagnosis for high-speed railway [83]	DBN
	Feature learning model for rotating machines [96]	CNN

(Continued)

TABLE 6.1 (*Continued*)

IoT Application Area	Proposed Model	Deep Learning Approach Used
Smart agriculture	Plant disease detection [101]	CNN
	Weed detection [102]	CNN
	Plant classification [103]	ALexNet
	Unsupervised cycle generative adversarial network [100]	GAN
	Soil monitoring [110]	WSN
	Hybrid Deep Learning predictor [72]	GRU
	DLEFN [114]	Fog Computing
	DenseNet [115]	DNN

6.4 CONCLUSION AND FUTURE ASPECTS

This chapter scrutinized the combined opportunities brought by Deep Learning in IoT and their effectiveness in results. IoT has provided us comfort in many application areas such as security, homes, agriculture, and industry. However, the minute features and obstacles to collect meaningful data have been given by Deep Learning approaches like CNN, RNN, auto-encoders, or GAN. Deep Learning has shown efficient results as compared to traditional approaches used and divulged complex features. The motivation behind this chapter is to contemplate the Deep Learning frameworks, and their contribution in providing effectiveness in results in IoT applications in all areas. However, the implementation of Deep Learning approaches has not been shown. More Deep Learning models can be proposed for future scope using existing and novel frameworks for more efficient and effective results to collect the complex features from data collected from different IoT applications.

The first and foremost challenge is the high velocity, volume, and variety of big IoT data. The '3Vs' of big data pose time and structure complexities in extracting useful features through the feature selection process. Since data classification problems are highly dependent on the number and quality of features selected for detection purposes, the scalability of data impacts the performance measures used to analyze data. Another challenge associated with voluminous data is the feasibility of training a Deep Learning model with central storage and processing devices. However, this problem may be resolved using distributed architecture, but further research is needed before making any claims.

Further, data from different sources may be noisy and unlabeled. Data needs to be cleaned and transformed before processing. Since Deep Learning models can analyze various IoT data collected from heterogeneous sources, handling different data formats is a challenging task for Deep Learning models. There is a lack of IoT datasets to test the proposed Deep Learning solutions and generate accurate results. This is one of the biggest challenges for integrating Deep Learning in IoT [4, 117].

REFERENCES

[1] E. Fleisch, "What Is the Internet of Things? An Economic Perspective," Auto-ID Labs White Pap. WP-BIZAPP-053, pp. 1–27, 2010

[2] M. A. Khan and K. Salah, "IoT Security: Review, Block Chain Solutions, and Open Challenges," *Future Generation Computer Systems*, Vol. 82, pp. 395–411, 2018.

[3] V. Mm, "Women's Safety Using IoT," *International Research Journal of Engineering and Technology (IRJET)*, Vol. 4, pp. 599–601, 2017.

[4] M. Mohammadi, "Deep Learning for IoT Big Data and Streaming Analytics: A Survey," *IEEE Communications Surveys & Tutorials*, Vol. 20, pp. 2923–2960, 2018.

[5] Qingchen Zhang, L. T. Yang, Z. Chen and P Li, "A Survey on Deep Learning for Big Data," *Information Fusion*, Vol. 42, pp. 146–157, 2018.

[6] Jie Tange, Dawei Sun, Shaoshan Liu and Jean-Luc Gaudiot, "Enabling Deep Learning on IoT Devices," *Computer*, Vol. 50, pp. 92–96, 2017.

[7] Mingcong Song, Kan Zhong, Jiaqi Zhang, Yang Hu, Duo Liu, Weigong Zhang, Jing Wang and Tao Li, "*In-Situ AI: towards Autonomous and Incremental Deep Learning for IoT Systems,*" in *2018 IEEE International Symposium on High Performance Computer Architecture*, Vienna, Austria, pp. 92–103.

[8] B. K. Mishra, D. Thakker, S. Mazumdar, D. Neagu and S. Simpson, "*Using Deep Learning for IoT-Enabled Smart Camera: A Use Case of Flood Monitoring,*" in *Proceedings of the 2019 10th International Conference on Dependable Systems, Services and Technologies (DESSERT)*, pp. 5–7, 2019.

[9] T. J. Hazen, "Microsoft and Liebherr Collaborating on New Generation of Smart Refrigerators." Sep. 2016 [Online]. Available: http://blogs.technet.microsoft.com/machinelearning/2016/09/02/

[10] J. Wang, H. Ding, F. Azamian, B. Zhou, C. Iribarren, S. Molloi and P. Baldi, "Detecting Cardiovascular Disease from Mammograms with Deep Learning," *IEEE Transactions on Medical Imaging*, Vol. 36, pp. 1172–1181, 2017.

[11] Yilin Yan, Min Chen, Saad Sadiq and Mei-Ling Shyu, "Efficient Imbalanced Multimedia Concept Retrieval by Deep Learning on Spark Clusters," *International Journal of Multimedia Data Engineering and Management*, Vol. 8, pp. 1–20, 2017.

[12] Maryam M. Najafabadi, Flavio Villanustre, Taghi M. Khoshgoftaar, Naeem Seliya, Randall Wald and Edin Muharemagic, "Deep Learning Applications and Challenges in Big Data Analytics," *Journal of Big Data*, Vol. 2, pp. 1–21, 2015.

[13] Samira Pouyanfar, Saad Sadiq, Yilin Yan, Haiman Tian, Yudong Tao, Maria Presa Reyes, Mei-Ling Shyu, Shu-Ching Chen and S. S. Iyengar, "Survey on Deep Learning: Algorithms, Techniques, and Applications," *ACM Comput. Surv.*, Vol. 51, pp. 1–36, Sep. 2018. https://doi.org/10.1145/3234150

[14] Quoc V. Le, "*Building High-Level Features Using Large Scale Unsupervised Learning,*" in *IEEE International Conference on Acoustics, Speech and Signal Processing*, pp. 8595–8598, May 2013.

[15] Alec Radford, Luke Metz and Soumith Chintala, "Unsupervised Representation Learning with Deep Convolutional Generative Adversarial Networks," *CoRR*, abs/1511.06434, 2015. Retrieved from http://arxiv.org/abs/1511.06434.

[16] L. Huang, J. Yin, B. Chen and S. Ye, "*Towards Unsupervised Single Image Dehazing With Deep Learning,*" in *2019 IEEE International Conference on Image Processing (ICIP)*, Taipei, Taiwan, pp. 2741–2745, 2019. doi: 10.1109/ICIP.2019.8803316.

[17] Doyen Sahoo, Quang Pham, Jing Lu and Steven C. H. Hoi, "*Online Deep Learning: Learning Deep Neural Networks on the Fly,*" in *Proceedings of the Twenty-Seventh International Joint Conference on Artificial Intelligence IJCAI 2018*, Stockholm, pp. 2660–2666, July 13–19.

[18] Ilya Sutskever, James Martens, George E. Dahl and Geoffrey E. Hinto, "*On the Importance of Initialization and Momentum in Deep Learning,*" in *International Conference on Machine Learning*, Leeds, UK, pp. 1139–1147, 2013.

[19] John C. Duchi, Elad Hazan and Yoram Singer, "*Adaptive Sub Gradient Methods for Online Learning and Stochastic Optimization,*" in *Conference on Learning Theory*, Haifa,Israel, pp. 257–269, 2010.

[20] Matthew D. Zeiler. "ADADELTA: An Adaptive Learning Rate Method," *CoRR*, abs/1212.5701, 22 Dec. 2012. Retrieved from http://arxiv.org/abs/1212.5701.

[21] T. Dozat, "Incorporating Nesterov Momentum into Adam," 2016, pp. 1–4. http://cs229.stanford.edu/proj2015/054report.pdf.

[22] Hang Su and Haoyu Chen, "Experiments on Parallel Training of Deep Neural Network Using Model Averaging," *CoRR*, abs/1507.01239, Jul 2015. Retrieved from http://arxiv.org/abs/1507.01239.

[23] Omry Yadan, Keith Adams, Yaniv Taigman and Marc'Aurelio Ranzato, "Multi-GPU training of Conv Nets," *CoRR*, abs/1312.5853, 2013.

[24] A. Al-Fuqaha, M. Guizani, M. Mohammadi, M. Aledhari and M. Ayyash, "Internet of Things: A Survey on Enabling Technologies, Protocols, and Applications," *IEEE Communications Surveys & Tutorials*, Vol. 17, no. 4, pp. 2347–2376, 2015.

[25] R. Tiwari, N. Sharma, I. Kaushik, A. Tiwari and B. Bhushan, "*Evolution of IoT & Data Analytics Using Deep Learning,*" in *2019 International Conference on Computing, Communication, and Intelligent Systems (ICCCIS)*, Greater Noida, India, 2019, pp. 418–423. doi: 10.1109/ICCCIS48478.2019.8974481.

[26] Shuochao Yao, Yiran Zhao, Aston Zhang, Shaohan Hu, Huajie Shao, Chao Zhang, Lu Su and Tarek Abdelzaher, "Deep Learning for the Internet of Things," *Computer*, vol. 51, no. 5, pp. 32–41, May 2018, doi: 10.1109/MC.2018.2381131.

[27] Md. A. Amanullah, R. A. A. Habeeb, F. H. Nasaruddin, A. Gani, E. Ahmed, A. S. M. Nainar, N. Md. Akim and Md. Imran "Deep Learning and Big Data Technologies for IoT Security," *Computer Communications*, Vol. 151, pp. 495–517, 2020. doi: https://doi.org/10.1016/j.comcom.2020.01.016 Reference: COMCOM 6137

[28] Y. Kim, "*Convolutional Neural Networks for Sentence Classification,*" in *Proceedings of the 2014 Conference on Empirical Methods in Natural Language Processing, EMNLP*, Doha, Qatar, pp. 1746–1751, Oct. 2014. URL http://aclweb.org/anthology/D/D14/D14-1181.pdf

[29] Zhongru Ren, Haomin Wu, Qian Ning, Iftikhar Hussain and Bingcai Chen, "End-to-end Malware Detection for Android IoT Devices Using Deep Learning," *Adhoc Networks*, online, Vol. 101, 2020. doi: https://doi.org/10.1016/j.adhoc.2020.102098

[30] J. Qiu, S. Nepal, W. Luo, L. Pan, Y. Tai, J. Zhang and Y. Xiang, "*Data-Driven Android Malware Intelligence: A Survey,*" in *International Conference on Machine Learning for Cyber Security*, Springer, Cham, pp. 183–202, 2019. https://doi.org/10.1007/978-3-030-30619-9_14

[31] https://pixabay.com/illustrations/security-secure-locked-technology-2168233/

[32] https://pixabay.com/photos/smart-city-communication-network-4168483/

[33] T. H. Huang and H. Kao, "*R2-D2: Color-Inspired Convolutional Neural Network (CNN)-Based Android Malware Detections,*" in *IEEE International Conference on Big Data*, Seattle, WA, pp. 2633–2642, Dec. 2018. https://doi.org/10.1109/BigData.2018.8622324

[34] M. A. Al-Garadi, A. Mohamed, A. K. Al-Ali, X. Du, I. Ali and M. Guizani, "A Survey of Machine and Deep Learning Methods for Internet of Things (IoT) Security," *IEEE Communications Surveys & Tutorials*, Vol. 22, no. 3, pp. 1646–1685, 2020. doi: 10.1109/COMST.2020.2988293.

[35] D. Kaur, R. Chahar and J. Ashta, "*IoT Based Women Security: A Contemplation,*" in *2020 International Conference on Emerging Smart Computing and Informatics (ESCI)*, Pune, India, pp. 257–262, 2020. doi: 10.1109/ESCI48226.2020.9167584.

[36] E. Raff, J. Barker, J. Sylvester, R. Brandon, B. Catanzaro and C. K. Nicholas, "*Malware Detection by Eating a Whole EXE,*" in *The Workshops of the The Thirty-Second AAAI Conference on Artificial Intelligence*, New Orleans, Louisiana, pp. 268–276, Feb. 2–7, 2018. https://aaai.org/ocs/index.php/WS/AAAIW18/paper/view/16422.

[37] W. G. Wong, "Developers Discuss IOT Security and Platforms Trends," accessed on May 1, 2019 (Jul 2015). https://www.electronicdesign.com/embedded/developers-discuss-iot-security-and-platforms-trends

[38] Abebe Abeshu Diro and Naveen Chilamkurti, "Distributed Attack Detection Scheme Using Deep Learning Approach for Internet of Things," *Future Generation Computer Systems*, Vol. 82, pp. 761–768, 2018. https://doi.org/10.1016/j.future.2017.08.043.

[39] A. Ferdowsi and W. Saad, "*Deep Learning-Based Dynamic Watermarking for Secure Signal Authentication in the Internet of Things,*" in *2018 IEEE International Conference on Communications (ICC)*, Kansas City, MO, pp. 1–6, 2018. doi: 10.1109/ICC.2018.8422728.

[40] Z. Tian, C. Luo, J. Qiu, X. Du and M. Guizani, "A Distributed Deep Learning System for Web Attack Detection on Edge Devices," in *IEEE Transactions on Industrial Informatics*, Vol. 16, no. 3, pp. 1963–1971, Mar 2020. doi: 10.1109/TII.2019.2938778.

[41] H. Jafari, O. Omotere, D. Adesina, H. Wu and L. Qian, "*IoT Devices Fingerprinting Using Deep Learning,*" in *2018 IEEE Military Communications Conference (MILCOM)*, Los Angeles, CA, pp. 1–9, 2018, doi: 10.1109/MILCOM.2018.8599826.

[42] M. Roopak, G. Yun Tian and J. Chambers, "*Deep Learning Models for Cyber Security in IoT Networks,*" in *2019 IEEE 9th Annual Computing and Communication Workshop and Conference (CCWC)*, Las Vegas, NV, pp. 0452–0457, 2019. doi: 10.1109/ CCWC.2019.8666588.

[43] Y. E. Sagduyu, Y. Shi and T. Erpek, "*IoT Network Security from the Perspective of Adversarial Deep Learning,*" in *2019 16th Annual IEEE International Conference on Sensing, Communication, and Networking (SECON)*, Boston, MA, pp. 1–9, 2019. doi: 10.1109/SAHCN.2019.8824956.

[44] M. Z. Alom and T. M. Taha, "*Network Intrusion Detection for Cyber Security Using Unsupervised Deep Learning Approaches,*" in *2017 IEEE National Aerospace and Electronics Conference (NAECON)*, Dayton, OH, 2017, pp. 63–69. doi: 10.1109/ NAECON.2017.8268746.

[45] A. Abeshu and N. Chilamkurti, "Deep Learning: The Frontier for Distributed Attack Detection in Fog-to-Things Computing," *IEEE Communications Magazine*, Vol. 56, no. 2, pp. 169–175, Feb. 2018. doi: 10.1109/MCOM.2018.1700332.

[46] M. Chen, X. Shi, Y. Zhang, D. Wu and M. Guizani, "Deep Features Learning for Medical Image Analysis with Convolutional Autoencoder Neural Network," in *IEEE Transactions on Big Data*, Vol. 7, no. 4, pp. 750–758, 1 Sept. 2021, doi: 10.1109/TBDATA.2017.2717439.

[47] N. Y. Hammerla, S. Halloran and T. Plötz, "*Deep, Convolutional, and Recurrent Models for Human Activity Recognition Using Wearables,*" in *Proceedings of 25th International Joint Conference Artificial Intelligence (IJCAI)* New York, NY, April 2016.

[48] Y. Gao, X. Xiang, N. Xiong, B. Huang, H. J. Lee, R. Alrifai, X. Jiang and Z. Fang, "Human Action Monitoring for Healthcare Based on Deep Learning," *IEEE Access*, Vol. 6, pp. 52277–52285, 2018.

[49] A. Prasoon, K. Petersen, C. Igel, F. Lauze, E. Dam and M. Nielsen, "*Deep Feature Learning for Knee Cartilage Segmentation Using a Triplanar Convolutional Neural Network,*" in *International Conference on Medical Image Computing and Computer-Assisted Intervention. (MICCAI)*, Springer, Berlin, Heidelberg. pp. 246–253, 2013.

[50] M. Chen, Y. Zhang, M. Qiu, N. Guizani and Y. Hao, "SPHA: Smart Personal Health Advisor Based on Deep Analytics," *IEEE Commun. Mag.*, Vol. 56, no. 3, pp. 164–169, Mar. 2018.

[51] https://www.sdsolutionsllc.com/implementation-of-occupational-safety-and-health-information-system/

[52] http://wallpaperswide.com/doctor_office-wallpapers.html

[53] X. Zeng, K. Cao and M. Zhang, *"Mobile Deep Pill: A Small-Footprint Mobile Deep Learning System for Recognizing Unconstrained Pill Images,"* in *Proceedings of 15th Annual International Conference Mobile System, Application, Services (Mobi Sys)* Niagara Falls, NY., pp. 56–67, 2017.

[54] A. R. Lopez, X. Giro-I-Nieto, J. Burdick and O. Marques, *"Skin Lesion Classification from Dermoscopic Images Using Deep Learning Techniques,"* in *Proceedings of 13th IASTED International Conference Biomedical Engineering (Bio Med)* Innsbruck, Austria , pp. 49–54, 2017.

[55] W.-J. Chang, L.-B. Chen, C.-H. Hsu, C.-P. Lin and T.-C. Yang, "A Deep Learning-Based Intelligent Medicine Recognition System for Chronic Patients," *IEEE Access*, Vol. 7, pp. 44441–44458, 2019.

[56] Shreshth Tuli, Nipam Basumatary, Sukhpal Singh Gill, Mohsen Kahani, Rajesh Chand Arya, Gurpreet Singh Wander and Rajkumar Buyya, "HealthFog: An Ensemble Deep Learning Based Smart Healthcare System for Automatic Diagnosis of Heart Diseases in integrated IoT and Fog Computing Environments," *Future Generation Computer Systems*, Vol. 104, pp. 187–200, 2020. https://doi.org/10.1016/j.future.2019.10.043.

[57] I. Azimi, J. Takalo-Mattila, A. Anzanpour, A. M. Rahmani, J. P. Soininen and P. Liljeberg., *"Empowering Healthcare IoT Systems with Hierarchical Edge-Based Deep Learning,"* in *Proceedings of the 2018 IEEE/ACM International Conference on Connected Health: Applications*, Washington, DC, Systems and Engineering Technologies, pp. 63–68, Sep. 2018.

[58] Farman Ali, Shaker El-Sappagh, S. M. Riazul Islam, Daehan Kwak, Amjad Ali, Muhammad Imran, Kyung-Sup Kwak, "A Smart Healthcare Monitoring System for Heart Disease Prediction Based on Ensemble Deep Learning and Feature Fusion," *Information Fusion*, Vol. 63, pp. 208–222, 2020. ISSN 1566-2535, https://doi.org/10.1016/j.inffus.2020.06.008

[59] M. Alhussein and G. Muhammad, "Automatic Voice Pathology Monitoring Using Parallel Deep Models for Smart Healthcare," *IEEE Access*, Vol. 7, pp. 46474–46479, 2019. doi: 10.1109/ACCESS.2019.2905597.

[60] P. Sundaravadivel, K. Kesavan, L. Kesavan, S. P. Mohanty and E. Kougianos, "Smart-Log: A Deep-Learning Based Automated Nutrition Monitoring System in the IoT," *IEEE Transactions on Consumer Electronics*, Vol. 64, no. 3, pp. 390–398, Aug. 2018. doi: 10.1109/TCE.2018.2867802.

[61] K. Muhammad, S. Khan, J. D. Ser and V. H. C. de Albuquerque, "Deep Learning for Multigrade Brain Tumor Classification in Smart Healthcare Systems: A Prospective Survey," *IEEE Transactions on Neural Networks and Learning Systems*, Vol. 32, no. 2, pp. 507–522, Jun 30.2020. doi: 10.1109/TNNLS.2020.2995800.

[62] M. A. Khan, M. S. Sarfraz, M. Alhaisoni, A. A. Albesher, S. Wang and I. Ashraf, "StomachNet: Optimal Deep Learning Features Fusion for Stomach Abnormalities Classification," *IEEE Access*, Vol. 8, pp. 197969–197981, 2020. doi: 10.1109/ACCESS.2020.3034217.

[63] A. Gumaei, M. M. Hassan, A. Alelaiwi and H. Alsalman, "A Hybrid Deep Learning Model for Human Activity Recognition Using Multimodal Body Sensing Data," *IEEE Access*, Vol. 7, pp. 99152–99160, 2019. doi: 10.1109/ACCESS.2019.2927134.

[64] M. H. Cılasun and H. Yalçın, "*A Deep Learning Approach to EEG Based Epilepsy Seizure Determination,*" in *2016 24th Signal Processing and Communication Application Conference (SIU)*, Zonguldak, pp. 1573–1576, 2016. doi: 10.1109/SIU.2016.7496054.

[65] Andrés Ortiz, Jorge Munilla, Juan M. Górriz and Javier Ramírez, "Ensembles of Deep Learning Architectures for the Early Diagnosis of the Alzheimer's Disease," *International Journal of Neural Systems*, Vol. 26, no. 07, 2016. https://doi.org/10.1142/S0129065716500258

[66] S. U. Amin, M. S. Hossain, G. Muhammad, M. Alhussein and M. A. Rahman, "Cognitive Smart Healthcare for Pathology Detection and Monitoring," *IEEE Access*, Vol. 7, pp. 10745–10753, 2019. doi: 10.1109/ACCESS.2019.2891390.

[67] Xiaoqiang Ma, Tai Yao, Menglan Hu, Yan Dong, Wei Liu, Fangxin Wang and Jiangchuan Liu, "A Survey on Deep Learning Empowered IoT Applications," *IEEE Access*, pp. 181721–181732, Dec. 2019. doi: 10.1109/ACCESS.2019.2958962

[68] Karen Quatromoni, "The Industrial Internet Consortium Announces Learnings from Its AI-Focused Testbed on a Deep Learning Facility," Toshiba: Press-Release. 2016. [Online]. Dec. 2018, Available: https://www.toshiba.co.jp/about/press/2016/10/pr1702.htm

[69] Y. Gu, Y. Chen, J. Liu and X. Jiang, "Semi-Supervised Deep Extreme Learning Machine for Wi-Fi Based Localization," *Neurocomputing*, Vol. 166, pp. 282–293, Oct. 2015.

[70] M. Mohammadi, A. Al-Fuqaha, M. Guizani and J.-S. Oh, "Semi-Supervised Deep Reinforcement Learning in Support of IoT and Smart City Services," *IEEE Internet Things Journal*, Vol. 5, pp. 624–635, April 2018.

[71] B. A. Erol, A. Majumdar, J. Lwowski, P. Benavidez, P. Rad and M. Jamshidi, "Improved Deep Neural Network Object Tracking System for Applications in Home Robotics," *Computational Intelligence for Pattern Recognition (Studies in Computational Intelligence)*, Vol. 777, Springer, Cham,2018. [Online]. Available: https://link.springer.com/chapter/10.1007/978-3-319-89629-8_14)

[72] S. Levine, P. Pastor, A. Krizhevsky, J. Ibarz and D. Quillen, "Learning Hand-Eye Coordination for Robotic Grasping with Deep Learning and Large-Scale Data Collection," *The International Journal of Robotics Research*, Vol. 37, pp. 421–436, 2018.

[73] L. Deng, "Artificial Intelligence in the Rising Wave of Deep Learning: The Historical Path and Future Outlook [Perspectives]," *IEEE Signal Processing Magazine*, Vol. 35, pp. 177–180, Jan. 2018.

[74] Faisal Mehmood, Israr Ullah, Shabir Ahmad and DoHyeun Kim, "Object Detection Mechanism Based on Deep Learning Algorithm Using Embedded IoT Devices for Smart Home Appliances Control in CoT," *Journal of Ambient Intelligence and Humanized Computing*, pp. 1–17, 2019. https://doi.org/10.1007/s12652-019-01272-8

[75] https://www.pinterest.com/a0939680390/building/

[76] V. Bianchi, M. Bassoli, G. Lombardo, P. Fornacciari, M. Mordonini and I. De Munari, "IoT Wearable Sensor and Deep Learning: An Integrated Approach for Personalized Human Activity Recognition in a Smart Home Environment," *IEEE Internet of Things Journal*, Vol. 6, no. 5, pp. 8553–8562, Oct. 2019. doi: 10.1109/JIOT.2019.2920283.

[77] Fotios Zantalis, Grigorios Koulouras, Sotiris Karabetsos and Dionisis Kandris, "A Review of Machine Learning and IoT in Smart Transportation," *Future Internet*, Vol. 11, 2019. doi:10.3390/fi11040094

[78] W. Huang, G. Song, H. Hong and K. Xie, "Deep Architecture for Traffic Flow Prediction: Deep Belief Networks with Multitask Learning," *IEEE Transaction on Intelligent Transportation System* Vol. 15, pp. 2191–2201, Apr. 2014.

[79] Z. Zhao, W. Chen, X. Wu, P. C. Y. Chen and J. Liu, "LSTM Network: A Deep Learning Approach for Short-Term Traffic Forecast," *IET Intelligent Transport Systems*, Vol. 11, pp. 68–75, 2017.

[80] S. Ren, K. He, R. Girshick and J. Sun, *"Faster R-CNN: Towards Real Time Object Detection with Region Proposal Networks,"* in *Proceedings of 28th International Conference on Neural Information Processing System (NIPS)*, Montreal, Canada, 2015.

[81] J. Redmon, S. Divvala, R. Girshick and A. Farhadi, *"You Only Look Once: United, Real-Time Object Detection,"* in *Proceedings of IEEE Conference on Computer Vision and Pattern Recognition (CVPR)*,Las Vegas, NV, pp. 779–788, Jun. 2016.

[82] W. Liu, D. Anguelov, D. Erhan, C. Szegedy, S. Reed, C.-Y. Fu and A. C. Berg, SSD: Single Shot MultiBox Detector. In: Leibe B., Matas J., Sebe N., Welling M. (eds) *Computer Vision – ECCV 2016*. Amsterdam, Netherlands. Lecture Notes in Computer Science, vol 9905. Springer, Cham. https://doi.org/10.1007/978-3-319-46448-0_2

[83] Jiateng Yin and Wentian Zhao, "Fault Diagnosis Network Design for Vehicle on-Board Equipments of High-Speed Railway: A Deep Learning Approach," *Engineering Applications of Artificial Intelligence*, Vol. 56, pp. 250–259, 2016. https://doi.org/10.1016/j.engappai.2016.10.002.

[84] A. Ferdowsi, U. Challita and W. Saad, "Deep Learning for Reliable Mobile Edge Analytics in Intelligent Transportation Systems: An Overview," *IEEE Vehicular Technology Magazine*, Vol. 14, no. 1, pp. 62–70, Mar. 2019. doi: 10.1109/MVT.2018.2883777.

[85] Jianqing Zhu; Jingchang Huang, Huanqiang Zeng; Xiaoqing Ye, Baoqing Li, Zhen Lei and Lixin Zheng, "Object Reidentification via Joint Quadruple Decorrelation Directional Deep Networks in Smart Transportation," *IEEE Internet of Things Journal*, Vol. 7, no. 4, pp. 2944–2954, Apr. 2020, doi: 10.1109/JIOT.2020.2963996.

[86] J. Ashraf, A. D. Bakhshi, N. Moustafa, H. Khurshid, A. Javed and A. Beheshti, "Novel Deep Learning-Enabled LSTM Autoencoder Architecture for Discovering Anomalous Events From Intelligent Transportation Systems," *IEEE Transactions on Intelligent Transportation Systems*, Vol. 22, no. 7, pp. 4507–4518, July 2021, doi: 10.1109/TITS.2020.3017882.

[87] Y. Zhang and T. Cheng, "A Deep Learning Approach to Infer Employment Status of Passengers by Using Smart Card Data," *IEEE Transactions on Intelligent Transportation Systems*, Vol. 21, no. 2, pp. 617–629, Feb. 2020. doi: 10.1109/TITS.2019.2896460.

[88] Z. Zhou, H. Liao, B. Gu, K. M. S. Huq, S. Mumtaz and J. Rodriguez, "Robust Mobile Crowd Sensing: When Deep Learning Meets Edge computing," *IEEE Network*, Vol. 32, no. 4, pp. 54–60, July/August 2018. doi: 10.1109/MNET.2018.1700442.

[89] A. Pandey, M. Puri and A. Varde, *"Object Detection with Neural Models, Deep Learning and Common Sense to Aid Smart Mobility,"* in *2018 IEEE 30th International Conference on Tools with Artificial Intelligence (ICTAI)*, Volos, pp. 859–863, 2018. doi: 10.1109/ICTAI.2018.00134.

[90] Kunja Bihari Swain, G. Santamanyu and Amiya Ranjan Senapati, *"Smart Industry Pollution Monitoring and Controlling Using LabVIEW Based IoT,"* in *2017 Third International Conference on Sensing, Signal Processing and Security (ICSSS)*, Chennai, India, pp. 74–78. IEEE.

[91] L. Li, K. Ota and M. Dong, "Deep Learning for Smart Industry: Efficient Manufacture Inspection System with Fog Computing," *IEEE Transaction on Industrial Informatics*, Vol. 14, no. 10, pp. 4665–4673, Oct. 2018.

[92] J. K. Park, B. K. Kwon, J. H. Park and D. J. Kang, "Machine Learning Based Imaging System for Surface Defect Inspection," *International Journal of Precision Engineering and Manufacturing-Green Technology*, Vol. 3, no. 3, pp. 303–310, Jul. 2016.

[93] J. Wang, J. Zhuang, L. Duan and W. Cheng, "*A Multi-Scale Convolution Neural Network for Featureless Fault Diagnosis*," in *2016 International Symposium on Flexible Automation (ISFA)*, Ohio, USA, IEEE, pp. 65–70, 2016.

[94] L. Wang, X. Zhao, J. Pei and G. Tang, "Transformer Fault Diagnosis Using Continuous Sparse Autoencoder," *Springer Plus*, Vol. 5, no. 1, pp. 1–13, 2016.

[95] Jay Lee, Hung-An Kao and Shanhu Yang, "Service Innovation and Smart Analytics for Industry 4.0 and Big Data Environment," *Procedia Cirp*, Vol. 16, pp. 3–8, 2014. doi: 10.1016/j.procir.2014.02.001

[96] Olivier Janssens, Viktor Slavkovikj, Bram Vervisch, Kurt Stockman, Mia Loccufier, Steven Verstockt, Rik Van de Walle and Sofie Van Hoecke, "Convolutional Neural Network Based Fault Detection for Rotating Machinery," *Journal of Sound and Vibration*, Vol. 377, pp. 331–345, 2016. https://doi.org/10.1016/j.jsv.2016.05.027.

[97] Ren R, Hung T and Tan KC, "A Generic Deep-Learning-Based Approach for Automated Surface Inspection," *IEEE Transactions on Cybernetics*, Vol. 48, no. 3, pp. 929–940, 2017.

[98] Fanyu Bu and Xin Wang, "A Smart Agriculture IoT System Based on Deep Reinforcement Learning,"*Future Generation Computer Systems*, Vol. 99, pp. 500–507, 2019. https://doi.org/10.1016/j.future.2019.04.041.

[99] V. S. Magomadov, "Deep Learning and Its Role in Smart Agriculture," *Journal of Physics: Conference Series*, Vol. 1399, no. 4, 2019. doi:10.1088/1742-6596/1399/4/044109.

[100] Nanyang Zhu, Xu Liu, Ziqian Liu, Kai Hu, Yingkuan Wang, Jinglu Tan, Min Huang, Qibing Zhu, Xunsheng Ji, Yongnian Jiang and Ya Guo "Deep Learning for Smart Agriculture: Concepts, Tools, Applications, and Opportunities," *International Journal of Agricultural and Biological Engineering*, Vol. 11, no. 4, pp. 32–44, 2018.

[101] Hulya Yalcin and Salar Razavi, "*Plant Classification Using Convolutional Neural Networks*," in *2016 Fifth International Conference on Agro-Geoinformatics (Agro-Geoinformatics)*, Tiajin,China, IEEE, pp. 1–5, 2016.

[102] J. L. Tang, D. Wang, Z. G. Zhang, L. J. He, Xin J and Xu Y, "Weed Identification Based on K-Means Feature Learning Combined with Convolutional Neural Network," *Computers and Electronics in Agriculture*, Vol. 135, pp. 63–70, 2017.

[103] Hulya Yalcin, "*Plant Phenology Recognition Using Deep Learning*," in *2017 6th International Conference on Agro-Geoinformatics*, Fairfax, VA, IEEE, pp. 1–5, 2017.

[104] S. Pandey, A. Solanki, "Music Instrument Recognition Using Deep Convolutional Neural Networks." *International Journal of Information Technology* 13(3), 129–149 (2019)

[105] R. Rajput, A. Solanki, "*Real-Time Analysis of Tweets Using Machine Learning and Semantic Analysis.*" in *International Conference on Communication and Computing Systems (ICCCS 2016)*, Taylor & Francis, Gurgaon, vol 138 issue 25, pp. 687–692, 9–11 Sept, 2016.

[106] R. Ahuja, A. Solanki: "*Movie Recommender System Using K-Means Clustering and K-Nearest Neighbor.*" in *Confuence-2019: 9th International Conference on Cloud Computing, Data Science & Engineering*, Amity University, Noida, vol. 1231, no. 21, pp. 25–38, 2019.

[107] A. Tayal, U. Kose, A. Solanki, A. Nayyar, J. A. M. Saucedo: "Efficiency Analysis for Stochastic Dynamic Facility Layout Problem Using Meta-Heuristic, Data Envelopment Analysis and Machine Learning," *Comput. Intell.* 36(1), 172–202, 2020.

[108] T. Singh, A. Nayyar, A. Solanki, "Multilingual Opinion Mining Movie Recommendation System Using RNN," in Singh, P., Pawłowski, W., Tanwar, S., Kumar, N., Rodrigues, J., Obaidat, M. (eds.) *Proceedings of First International Conference on Computing, Communications, and Cyber-Security (IC4S 2019)*. Lecture Notes in Networks and Systems, vol 121. Springer, Singapore, 2020. https://doi.org/https://doi.org/10.1007/978-981-15-3369-3_44.

[109] Ruud Barth, Joris Jsselmuiden, Jochen Hemming and Eldert J. van Henten, "*Optimising Realism of Synthetic Agricultural Images Using Cycle Generative Adversarial Networks*," in *Proceedings of the IEEE IROS workshop on Agricultural Robotics*, pp. 18–22, 2017. http://library.wur.nl/WebQuery/wurpubs/533105.

[110] S. Aruul Mozhi Varman, A. R. Baskaran, S. Aravindh and E. Prabhu, "*Deep Learning and IoT for Smart Agriculture Using WSN*," in *2017 IEEE International Conference on Computational Intelligence and Computing Research (ICCIC)*, Coimbatore, pp. 1–6, 2017. doi: 10.1109/ICCIC.2017.8524140.

[111] Xue-Bo Jin, Nian-Xiang Yang, Xiao-Yi Wang, Yu-Ting Bai, Ting-Li Su and Jian-Lei Kong, "Hybrid Deep Learning Predictor for Smart Agriculture Sensing Based on Empirical Mode Decomposition and Gated Recurrent Unit Group Model," *Sensors*, Vol. 20, pp. 1–20, 2020.

[112] M. Roopaei, P. Rad and K. R. Choo, "Cloud of Things in Smart Agriculture: Intelligent Irrigation Monitoring by Thermal Imaging," *IEEE Cloud Computing*, Vol. 4, no. 1, pp. 10–15, Jan–Feb. 2017, doi: 10.1109/MCC.2017.5.

[113] https://iotdesignpro.com/articles/smart-farming-iot-applications-in-agriculture

[114] Kyuchang Lee, Bhagya Nathali Silva and Kijun Han, "Deep Learning Entrusted to Fog Nodes (DLEFN) Based Smart Agriculture," *Applied Sciences*, Vol. 10, 2020.

[115] L. Ale, A. Sheta, L. Li, Y. Wang and N. Zhang, "*Deep Learning Based Plant Disease Detection for Smart Agriculture*," *2019 IEEE Globecom Workshops (GC Wkshps)*, Waikoloa, HI, pp. 1–6, 2019, doi: 10.1109/GCWkshps45667.2019.9024439.

[116] Ahmed Dawoud, Seyed Shahristani and Chun Raun, "Deep Learning and Software-Defined Networks: Towards Secure IoT Architecture," *Internet of Things*, Vol. 3, pp. 82–89, 2018. https://doi.org/10.1016/j.iot.2018.09.003.

[117] Xue-Wen Chen, and Xiaotong Lin, "Big Data Deep Learning: Challenges and Perspectives," *IEEE Access*, Vol. 2, pp. 514–525, 2014.

7 Prediction and Classification Analysis of Type-2 Diabetes Using Machine Learning Approaches

Ritu Aggarwal
Maharishi Markandeshwar Institute of Computer
Technology & Business Management, Haryana, India

Zdzislaw Polkowski
Jan Wyzykowski University, Polkowicw, Poland

CONTENTS

7.1 INTRODUCTION

Diabetes is a heterogeneous grouping of muddles which is called a hereditary disorder. This results in an excessive amount of glucose in the blood. The main reason for diabetes is genetics. Some diabetes patients take the meal in large amounts and eat less [1, 2]. Both of these results in the development of diabetes, another name

DOI: 10.1201/9781003156123-7

of diabetes is diabetes mellitus [3]. The hemoglobin glucose levels are higher than an interlude that is developing metabolic-related issues. The syndrome of diabetes is excessive urination discrete at frequent intervals, it may lead to death, or life-threatening issues like brain stroke, ulcers, kidney disease, foot injury [4–6], etc. if the body is not producing enough tissues that are required for body and cell tissues that lead to diabetes. Diabetes disease exists in three phases [7].

Category-1 diabetes is accumulated by the giblets, which generate less insulin according to the body's requirement [8–10]. This condition is called insulin-assistant diabetes. The treatment is done by giving external insulin to the body to make up for the less insulin produced by giblets [11].

Category-2 diabetes is distinct when body insulin resists the body cells then reacts differently as to the normal insulin. It means that no insulin in the body is called non-insulin-dependent diabetes [12–14]. It commonly developing with high BMI found in people, leading to an inactive lifestyle. Another type of diabetes is gestational diabetes which is observed at the time of pregnancy. The normal ranges of glucose levels in the human body are 70–99 mg/dl. Two ways of [15, 16] identifying diabetes, fasting, and non-fasting sugar [17]. If an individual is having more than 126 mg/dl of fasting glucose level, in medical [17, 18] terms if a person has glucose quantity or deliberation in blood ranges from 100 to 125 mg/dl then the person can be called a diabetic patient [19, 20].

This research aims to prognoses diabetes in the early stages and determines [21, 22] its risks and development by using different machine learning classifier. In Section 7.2 (Background), the works by the existing researchers are mentioned. In Section 7.3 (The Different Prediction Models of Machine Learning), the various classifications and presaging algorithms of ML are discussed. In Section 7.4 (Methodology and Model Diagram of the Proposed Work), the dataset for diabetes used in this study is shown. Section 7.5 (Results and Their Experiments) presents the experiments and the results, and in Section 7.6 (Conclusion and Future Scope), the conclusion and future scope are summarized.

7.2   BACKGROUND

Previous researchers' works are discussed in this Background. Many researchers have used PIMA Datasets for predicting the disease. This chapter used the survey dataset and PIMA dataset. Swapna et al. [23] proposed electrocardiogram (ECG) signals for the prediction of diabetes by using different deep learning methods. It extracts the best feature of the Support Vector Machine (SVM) by using convolution neural networks. Sisodia et al. [24] in their work applied machine learning algorithms on the PIDD datasets for the DT, NB, and SVM. The Naïve Bayes has the best accuracy, i.e. 76.03%. Han Wu et al. [25] used the data mining techniques for better precision and prediction for Category-2 diabetes by taking a sample of individuals. The risk prediction is 95.42%. The researchers were selected many instances from datasets and 100. Meng et al. [26] proposed a methodology comparing different ML algorithms like LR, ANN, and DT used to diagnose diabetes. It has taken 12 risk factors that include many attributes like diabetes, trees of work, BMI, age, gender, marital status, coffee drinking, salty foods, physical activity, fish consumption, etc. The decision tree gives more accuracy for this model. Choubey et al. [27] used the hybrid approach with GA. It implements it by the RBFNN. This Chapter proposed a feature selection on the RBFNN using GA for classification. Tigga et al. in [28] projected a technique used for diabetic prediction by using PIDD datasets and selected features to check the risk of diabetes by the most prominent attributes such as the no. of pregnancies, BMI, and glucose quantity. Huang et al. [29] did the classification and feature selection by applying the machine learning classifiers. It has taken the different variables used in their studies, such as age, diet, insulin, and blood sugar. Tigga and Garg [30] proposed a methodology for predicting diabetes by means of the ML taxonomy. It used the 18 attributes in their research by the questionnaire on datasets and PIM datasets. The best meticulousness is finding with the help of the Random Forest that is 94.10%.

7.3 THE DIFFERENT PREDICTION MODELS OF MACHINE LEARNING

This section describes the various classifications and presaging algorithms of ML. With the help of these algorithms, performance and exactness are achieved. Following are the classifications and presaging algorithms [31, 32]:

7.3.1 SVM

It is a category of supervised learning. It is based on the taxonomy model. It employs two classes for training the sample [33–35]. It separates the highest margin that is called a Hyperplane [36]. A Hyperplane consists of two classes. Based on the accuracy, the best Hyperplane is selected to show which class is better. Mathematically, it is to calculate by this equation [37, 38].

$$wTX + B = -1 \tag{7.1}$$

It is used to maximize the distance between the Hyperplane [39, 40].

$$wTX + B = 1 \tag{7.2}$$

By the above equations (7.1 and 7.2), the Hyperplane is defined [41, 42].

7.3.2 NAÏVE BAYES

It defines all features, either is independent or dependent, or unrelated. It is employed for classification techniques [24]. The naïve only defines the particular feature of the class. The feature selection is based on conditional probability [43, 44]. It balances the data and their missing values [7].

7.3.3 LOGISTIC REGRESSION

It is based on set which are dependent upon the discrete values. This discrete value is used to split the observations [45, 46]. The output for this classifier is based on the opportunity feature [47–51]. Sigma and characteristic function is used to check the values between 0 and 1 [47].

7.3.4 K-NN

This algorithm is based on the same set of rules that sorted the complete dataset [26]. It has taken two classes. it used for classifying near values and once the data classify the neighbors of that [52, 53] element are selected based on the votes of the label [24].

7.3.5 Random Forest

RF is a type of supervised learning approach. It uses both classification and regression [54, 55]. This classifier creates multiple DTs to train the data [56]. The votes choose the final aggregation of different decision trees for their class of test objects [57, 58].

7.4 METHODOLOGY AND MODEL DIAGRAM OF THE PROPOSED WORK

This projected work has taken the dataset from UCI/ Kaggle for the PIMA Database. It consists of 768 instances. In this research comparison and accuracy are found using five machine learning algorithms. For accuracy consistency, the 10-fold cross-validation method is used and Table 7.1 shows the dataset for diabetes used in this study.

With the help of a dataset of diabetes, the attributes define with their input values, such as age, BMI, BP, pregnancies, etc., and Figure 7.1 shows the different attributes.

7.4.1 Describes the Methodology and Model Diagram for the Proposed Work

Jupyter notebook and python language are used for coding. Five different machine learning algorithms were used: Random Forest, LR, NB, K-NN, and SVM. This proposed framework took the data from the Pima dataset. By these datasets, we design a model diagram for this proposed work in Figure 7.2 Proposed work Methodology.

TABLE 7.1
Dataset for Diabetes

Sr. No	Abbreviations of Attribute	Attribute	Description
1.	PR	Pregnancies	In numbers
2.	GE	In hour Glucose	In an oral glucose tolerance test, the Plasma glucose deliberation 2 hours
3.	BP	Blood pressure	Diastolic blood pressure (mm Hg)
4.	ST	Skin Thickness in mm	Skinfold thickness Triceps
5.	IN	Insulin	2-hour serum insulin (mu U/ml)
6.	BMI	BMI	Body mass index (weight in kg/(height in nm))
7.	PDF	Pedigree Function Diabetes	Pedigree Function Diabetes
8.	AG	Age	21 or above
9.	TAR	Prediction result (Diabetes/non-diabetes)	Diabetic: 268 Non-diabetic:500

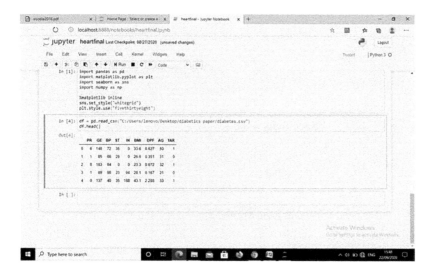

FIGURE 7.1 Different attributes or instances mentioned in Table 7.1 (i.e. Dataset for Diabetes).

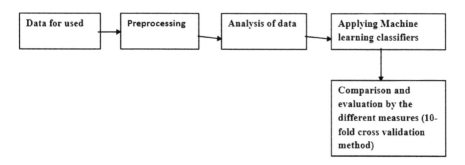

FIGURE 7.2 Proposed work methodology.

7.5 RESULTS AND THEIR EXPERIMENTS

Dataset is given above the figure and their results for this work is shown as, the diabetes patient represents value =1 and non_ diabetes patient=0. For train the dataset, the sample for the dataset is separated into equal parts for training and testing. By testing and training, the confusion matrix is produced. These matrices are estimated for the values of True_ N, True_ F, False_ N, and False. By using these accuracy measures the diabetes performance is calculated; the following given formula is used:

$$\text{Accuracy} = \frac{True_N + True\_P}{False\_p + False\_N + True_N + True\_p} \tag{7.3}$$

$$Error = \frac{False\_p + False\_N}{False\_p + False\_N + True\_N + True\_p} \tag{7.4}$$

$$Sensitivity = \frac{True\_P}{True\_P + False\_N} \tag{7.5}$$

$$Specificity = \frac{True\_N}{True\_N + False\_P} \tag{7.6}$$

$$Precision = \frac{True\_P}{True\_P + False\_P} \tag{7.7}$$

$$F\text{-Score} = 2*(precision*sensitivity/precision+sensitivity) \tag{7.8}$$

Table 7.2 shows the Comparative Analysis for the Different Classification Measures. Cross-validation is also used for testing this model. The performance of this model is checked by means of a 10-fold cross-validation method. By the value of k, the value is calculated for each dataset.

From the above results 10-fold cross-validation accuracy for Logistic Regression is very best and accurate. Precision is greater than 6 and the error is very less by applying the cross-validation method. The sensitivity is 0.682, and the F-measure score is greater than 0.760 (Figure 7.3).

Using different classification measures, the performance of each classifier is calculated and measured. Table 7.3 shows Different performance evaluation measures for each classifier separately. This graph shows the accuracy of classifier and checks which classifier has the highest accuracy score. As discussed earlier, the LR gives the best results when applying the cross-validation method.

By using a 10–fold cross-validation method logistic regression gives more accuracy than algorithms. Figure 7.4 shows the results of different performance evaluation measures for each classifier.

In the accuracy and the F-Score graph accuracy is 76 % for LR and F-score is also 76 % is also high when compared to it.

TABLE 7.2
Comparative Analysis for the Different Classification Measures

Classification Accuracy	LR	K-NN	SVM	NB	RF
Accuracy	0.76	0.71	0.72	0.74	0.75
Error	0.262	0.269	0.265	0.385	0.327
Sensitivity	0.789	0.738	0.765	0.72	0.769
Specificity	0.682	0.503	0.616	0.561	0.651
Precision	0.749	0.732	0.726	0.674	0.745
F-Score	0.761	0.734	0.611	0.583	0.743
10-fold	1.000	0.734	0.617	0.726	0.874

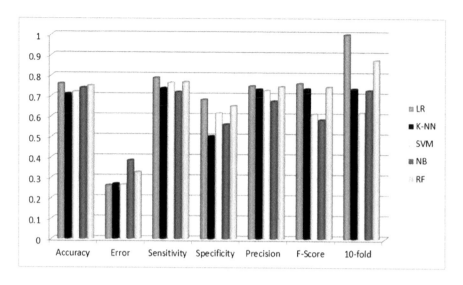

FIGURE 7.3   Comparisons between classifiers by different classification measures (by above Table 7.2).

TABLE 7.3
Different Performance Evaluation Measures for Each Classifier (A, B, C, D, E, F)

	LR	K-NN	SVM	NB	RF
A. Results of Accuracy					
Accuracy	0.76	0.71	0.72	0.74	0.75
Accuracy is 76 % for logistic regression as compared with other classification algorithms.					
B. Results of F-Score					
F-Score	0.76	0.734	0.611	0.583	0.743
F-score is 76 % for logistic regression as compared with other classification algorithms.					
C. Results of Specificity					
Specificity	0.682	0.503	0.616	0.561	0.651
Specificity is 68.2 % for logistic regression as compared with other classification algorithms.					
D. Results of Error					
Error	0.262	0.269	0.265	0.385	0.327
The error rate is 26.2 % for logistic regression is very less as compared with other classification algorithms.					
E. Results of Sensitivity					
Sensitivity	0.789	0.738	0.765	0.72	0.769
Sensitivity is 78.6 % for logistic regression as compared with other classification algorithms					
E. Results of 10-fold					
10-fold	1	0.734	0.617	0.726	0.874

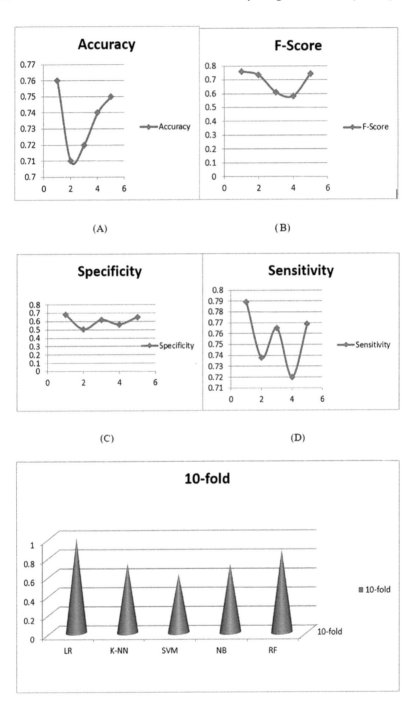

FIGURE 7.4   Results of different performance evaluation measures for each classifier (A, B, C, D, and E).

The specificity and sensitivity graph analysis shows the is best and high score as compared to other classifier.

With the help of 10-fold graph analyze the performance of LR and their results are calculated and achieved.

7.6 CONCLUSION AND FUTURE SCOPE

One of the major challenges in health care by using predictive analytics is very difficult to identify the risk of diabetes for its early stages. In this chapter, five popular machine learning classifier is used and implemented. The results are considered by taking the datasets from PIMA datasets. The dataset is taken PIMA datasets for 768 records. The algorithm of the classifier the logistic regression, gives the highest accuracy that is 76% for predicting diabetes. In future work for the best results, the parameter tuning model could be used for better accuracy and performance.

REFERENCES

[1] Kumari, V.A., Chitra, R., "Classification of diabetes disease using support vector machine", *International Journal of Engineering Research and Applications (IJERA)*, vol. 3, 2016, pp. 1797–1801. www.ijera.com.

[2] Kavakiotis, I., Tsave, O., Salifoglou, A., Maglaveras, N., Vlahavas, I., Chouvarda, I., "Machine learning and data mining methods in diabetes", *Research Computational and Structural Biotechnology Journal*, vol. 15, 2017, pp. 104–116.

[3] Eswari, T., Sampath, P., Lavanya, S. "Predictive methodology for diabetic data analysis in big data", *Procedia Computer Science*, 2015, pp. 203–208.

[4] Pradhan, P.M.A., Bamnote, G.R., Tribhuvan, V., Jadhav, K., Chabukswar, V., Dhobale, V, "A genetic programming approach for detection of diabetes", *International Journal of Computational Engineering Research*, vol. 2, 2012, pp. 91–94.

[5] Perveen, S., Shahbaz, M., Guergachi, A., Keshavjee, K., "Performance analysis of data mining classification techniques to predict diabetes", *Procedia Computer Science*, vol. 82, 2016, pp. 115–121.

[6] Sisodia, D., Shrivastava, S.K., Jain, R.C., "*ISVM for face recognition*", *International Conference on Computational Intelligence and Communication Networks, CICN*, vol. 2010, 2010, pp. 554–559.

[7] Tigga, N., Garg, S., "Prediction of Type 2 diabetes using machine learning classification methods", *Procedia Computer Science*, 2020, pp. 706–716.

[8] Pradhan, M., Bamnote, G.R., "Design of classifier for detection of diabetes mellitus using genetic programming", *Advances in Intelligent Systems and Computing*, vol. 1, 2014, pp. 763–770.

[9] Iyer, A., Jeyalatha, S., Sumbaly, R., "Diagnosis of diabetes using classification mining techniques", *International Journal of Data Mining & Knowledge Management Process*, vol. 5, 2015, pp. 1–14.

[10] Solanki, A., Pandey, S., "Music instrument recognition using deep convolutional neural networks", *Int. J. Inf. Technol. (IJITEE)*, 8, 2019, pp. 1076–1079.

[11] Orabi, K.M., Kamal, Y.M., Rabah, T. M., "*Early predictive system for diabetes mellitus disease*", *Industrial Conference on Data Mining, Springer*, 2016, pp. 420–427.

[12] Zou, Q., Qu, K., Luo, Y., Yin, D., Ju, Y., Tang, H., "Predicting diabetes mellitus with machine learning techniques", *Frontiers in Genetics*, 2018, pp. 515.

[13] Bhargavi, V. R., Senapati, R. K., "Curvelet fusion enhancement based evaluation of diabetic retinopathy by the identification of exudates in optic color fundus images", *Biomedical Engineering-Applications, Basis and Communications*, vol. 28, no. 6, 2016, ISSN 10162372.

[14] Kumar, K. V. V., Kishore, P. V. V., "Indian classical dance mudra classification using HOG features and SVM Classifier", *International Journal of Electrical and Computer Engineering*, 2016, pp. 2537–2546.

[15] Singh, T., Nayyar, A., Solanki, A., "*Multilingual opinion mining movie recommendation system using RNN*", *Proceedings of First International Conference on Computing, Communications, and Cyber-Security (IC4S 2019)*, 2020, pp. 589–605, Singapore. Springer Singapore.

[16] Ahuja, R., Solanki, A., Nayyar, A., "Movie recommender system using k-means clustering and k-nearest neighbor", 2019, pp. 263–268.

[17] Nai-Arun, N., Moungmai, R., "Comparison of classifiers for the risk of diabetes prediction", *Procedia Computer Science.*, 2015, pp. 132–142.

[18] Garner, S. R., "*Weka: The Waikato environment for knowledge analysis*", *Proceedings of the New Zealand Computer Science Research Students Conference*, Citeseer, 1995, pp. 57–64.

[19] Tayal, A., Kose, U., Solanki, A., Nayyar, A., Saucedo, J. A. M., "Efficiency analysis for stochastic dynamic facility layout problem using meta-heuristic, data envelopment analysis and machine learning", *Comput. Intell.*, vol. 36, no. 1, 2019, pp. 172–202.

[20] Tayal, A., Solanki, A., Singh, S. P., "Integrated frame work for identifying sustainable manufacturing layouts based on big data, machine learning, meta-heuristic and data envelopment analysis", *Sustain. Cities Soc.*, vol. 62, pp. 102383. https://doi.org/10.1016/j.scs.2020.10238.

[21] Mirza, S.S., Rahman, M.Z.U., "Efficient adaptive filtering techniques for thoracic electrical bioimpedance analysis in health care systems", *Journal of Medical Imaging and Health Informatics*, 2017, pp. 1126–1138.

[22] Rao, G. A., Syamala, K., Kishore, P. V. V., Sastry, A. S. C. S., "Deep convolutional neural networks for sign language recognition", *International Journal of Engineering and Technology (UAE)*, 2018,pp. 62–70.

[23] Kavakiotis, I., Tsave, O., Salifoglou, A., Maglaveras, N., Vlahavas, I., Chouvarda, I. "Machine learning and data mining methods in diabetes research", *Computational and Structural Biotechnology Journal*, 2017, pp. 104–116.

[24] Sisodia, D., Sisodia, D. S. "Prediction of diabetes using classification algorithms", *Procedia Computer Science*, 2018, pp. 1578–1585.

[25] Wu, H., YangS., Huang, Z., He, J., Wang, X. "Type 2 diabetes mellitus prediction model based on data mining", *Informatics in Medicine Unlocked*, pp. 100–107.

[26] Meng, X.H., Huang, Y.X., Rao, D.P., Zhang, Q., Liu, Q. "Comparison of three data mining models for predicting diabetes or pre diabetes by risk factors", *The Kaohsiung Journal of Medical Sciences*, 2013, pp. 93–99.

[27] Choubey, D.K., Paul, S. "GA_RBF NN: a classification system for diabetes", *International Journal of Biomedical Engineering and Technology*, pp. 71–93.

[28] Tigga, N. P., Garg, S. "Predicting type 2 diabetes using logistic regression", accepted to publish in *Lecture Notes of Electrical Engineering*, Springer. 2014, pp. 45.

[29] Huang, Y., McCullough, P., Black, N., Harper, R. "Feature selection and classification model construction on type 2 diabetic patients' data", *Artificial Intelligence*, 2007, pp. 251–262.

[30] Chiarelli, P. A., Hauptman, J. S., Browd, S. R., "Machine learning and the prediction of hydrocephalus," *JAMA Pediatric.*, vol. 172, no. 2, 2018, pp. 116.

[31] Patil, B. M., Joshi, R.C., Toshniwal, D., *"Association rule for classification of type-2 diabetic patients"*, *ICMLC '10 Proceedings of the 2010 Second International Conference on Machine Learning and Computing*, February 9–11, 2010.

[32] Khan, D. M., Mohamudally, N., "An integration of K-means and decision tree (ID3) towards a more efficient data mining algorithm", *Journal of Computing*, vol. 3, no. 12, December 2011. pp. 3415–3515.

[33] Reddy, S. S., Suman, M., Prakash, K. N., "Micro aneurysms detection using artificial neural networks", *Lecture Notes in Electrical Engineering*, 2018, pp. 273–282.

[34] Putluri, S., Ur Rahman, M. Z., Fathima, S. Y., "Cloud-based adaptive exon prediction for DNA analysis", *Lecture Notes in Electrical Engineering*, 2018, pp. 409–417.

[35] Priyam, A., Gupta, R., Rathee, A., Srivastava, S., "Comparative analysis of decision tree classification algorithms", *International Journal of Current Engineering and Technology*, vol. 3, 2013, pp. 334–337.

[36] https://www.kaggle.com/uciml/pima-indians-diabetes-database.

[37] Rajesh, K., Sangeetha, V., "Application of data mining methods and techniques for diabetes diagnosis", *International Journal of Engineering and Innovative Technology (IJEIT)*, vol. 2, no. 3, September 2012.

[38] Kahramanli, H., Allahverdi, N., "Design of a hybrid system for the diabetes and heart disease", *Expert Systems with Applications: An International Journal*, vol. 35, no. 1–2, July, 2008.

[39] Suresh Kumar, P., Pranavi, S., *"Performance analysis of machine learning algorithms on diabetes dataset using big data analytics"*, *International Conference on Infocom Technologies and Unmanned Systems*, 978-1-5386-0514-1, Dec. 2017, pp. 18–20.

[40] Butwall, M., Kumar, S., "A data mining approach for the diagnosis of diabetes mellitus using random forest classifier", *International Journal of Computer Applications*, vol. 120, no. 8, 2015.

[41] Saravanakumar, N. M., Eswari, T., Sampath, P., Lavanya, S., *"Predictive methodology for diabetic data analysis in big data"*, *2nd International Symposium on Big Data and Cloud Computing*, 2015.

[42] Aiswarya Iyer, S., Jeyalathaand, R. S., "Diagnosis of diabetes using classification mining techniques", *International Journal of Data Mining & Knowledge Management Process (IJDKP)*, vol. 5, no. 1, January 2015.

[43] Rajesh, V., Saikumar, K., Ahammad, S. K. H., "A telemedicine technology for cardiovascular patients diagnosis feature using knn-mpm algorithm", *Journal of International Pharmaceutical Research*, 2019, pp. 72.

[44] Nithya, B., Ilango, V., *"Predictive analytics in health care using machine learning tools and techniques"*, *International Conference on Intelligent Computing and Control Systems*, 2017. Saravana Kumar, N. M., Eswari, T., Sampath, P. and Lavanya, S., *"Predictive methodology for diabetic data analysis in big data"*, *2nd International Symposium on Big Data and Cloud Computing*, 2015.

[45] Kalyankar, Gauri D., Poojara, Shivananda R., Dharwadkar, Nagaraj V., *"Predictive analysis of diabetic patient data using machine learning and hadoop"*, *International Conference on I-SMAC*, 2017.

[46] Anand, A., Shakti, D., *"Prediction of diabetes based on personal lifestyle indicators"*, *1st International Conference on Next Generation Computing Technologies*, 2015.

[47] Swapna, G., VinayakumarR., SomanK. P., "Diabetes detection using deep learning algorithms", *ICT Express*, 2018, pp. 243–246.

[48] Jindal, A., Dua, A., Kumar, N., Das, A. K., Vasilakos, A. V., Rodrigues, J. J. P. C., "Providing healthcare-as-a-service using fuzzy rule-based big data analytics in cloud computing", *IEEE J. Biomed. Heal. Informatics*, 2018, pp. 1–1.

[49] Kumar, N. M. S., Eswari, T., Sampath, P., Lavanya, S., "Predictive methodology for diabetic data analysis in big data", *Procedia Comput. Sci.*, vol. 50, 2015, pp. 203–208.

[50] Hadoop, *International Conference on I-SMAC*, 978-1-5090-3243-3, 2017.

[51] Nithya, B., Ilango, V., *"Predictive analytics in health care using machine learning tools and techniques"*, International Conference on Intelligent Computing and Control Systems, 978-1-5386-2745-7, 2017.

[52] Cheng, L., Hammond, H., Ye, Z., Zhan, X., Dravid, G., "Human adult marrow cells support prolonged expansion of human embryonic stem cells in culture", *Stem Cells*, vol. 21, 2003, pp. 131–142.

[53] Schulz, T. C., Noggle, S. A., Palmarini, G. M., "Differentiation of human embryonic stem cells to dopaminergic neurons in serum-free suspension culture", *Stem Cells*, vol. 22, 2004, pp. 1218–1238.

[54] Dave, S. D., Vanikar, A. V., Trivedi, H. L., "Novel therapy for insulin-dependent diabetes mellitus: infusion of in vitro-generated insulin-secreting cells", *Clin. Exp. Med.*, vol. 15, no. 1, 2015, pp. 41–45.

[55] Guan, L. X., Guan, H., Li, H. B., "Therapeutic efficacy of umbilical cord-derived mesenchymal stem cells in patients with type 2 diabetes", *Exp. Ther. Med.*, vol. 9, no. 5, 2015, pp. 1623–1630.

[56] Baskar, M., Gnansekaran, T., "Multi model network analysis for improved intrusion tracing towards mitigating DDoS attack", *Asian Journal of Research in Social Sciences and Humanities*, vol. 7, no. 3, 2017, pp. 1343–1353.

[57] Han, Y. F., Sun, T. J., Han, Y. Q., "Clinical perspectives on mesenchymal stem cells promoting wound healing in diabetes mellitus patients by inducing autophagy", *Eur. Rev. Med. Pharmacol. Sci.*, vol. 19, no. 14, 2015, pp. 2666–2670.

[58] Baskar, M., Gnansekaran, T., "Developing efficient intrusion tracking system using region based traffic impact measure towards the denial of service attack mitigation", *Journal of Computational and Theoretical Nano science*, vol. 14, no. 7, 2017, pp. 3576–3582.

8 Internet of Thing-Based Monitoring Systems and Their Applications

N. Vijayalakshmi, S. Sindhu, and J. Dhilipan

SRM IST, Tamilnadu, India

CONTENTS

DOI: 10.1201/9781003156123-8

8.1 INTRODUCTION

The monitoring system had a major role in all the fields like agriculture, production, security issues, etc. Nowadays, all devices are connected under the circumstance of the Internet [1, 2]. The Internet makes the interactivity of the controlling system easier and eases its use. Nowadays, all environments are developed by science and techniques based on wireless-technology, which always senses data. The main important role of the entire digital device is to monitor the process [4]. Another fact is to keep the monitored data safe in this monitoring system, which may consume a shorter time for data analysis [3]. It reduces the time consumption for the maintenance process of every individual activity of the devices. They are accessing and monitoring the process of any environment from everywhere.

8.2 ADVANTAGES

This monitoring system is used to monitor the device or situation from everywhere. It doesn't require any specific location for the analyzing process [5]. It has an easy alert system. to easily analyze the huge volume of data, and it keeps the data into permanent storage and maintains the analyzed data [6, 7]. The main advantage is that it consumes lesser amount of time for decision-making, which provides higher security for the controlled devices.

8.3 APPLICATIONS IN IoT

There are many applications that are developed using IoT [8]. The IoT-based monitoring system is a health monitoring system that monitors the patient's body temperature and heartbeat rate, blood pressure, and other health-related symptoms. The child monitoring system/old age monitoring system monitors the kids and old age people's health conditions and track their location for security issues [11]. Most schools and colleges use the bus location tracking and the vehicle's fuel monitoring processfor finding the location and for safety measures. The power monitoring and power consumption monitoring systems may check current daily consumption and low power supply and other consumer activities [10]. The marine radioactivity-based monitoring system monitors any suspicious activities and security measures [12, 13]. And all other remaining monitoring systems are smart machine condition monitoring systems and android-based monitoring systems,etc.

8.4 GENERAL ARCHITECTURE OF THE SYSTEM

The sensor is important for all the control processing of the entire unit [14]. Sensors will sense the data and that data will be shared on to the cloud for data analysis. Based upon the output, the decision will be taken like alert sound or alert message, etc. (Figure 8.1).

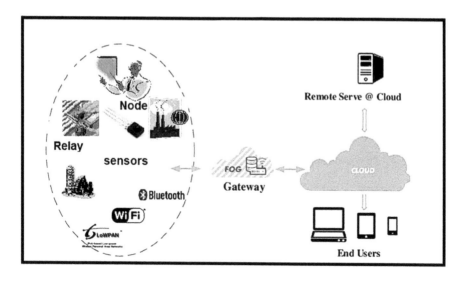

FIGURE 8.1   Common architecture for IoT-based health system.

The gateway is used to create the interaction between the device and the cloud memory and all other connected devices [16, 17]. The RFID tags transmit the sensed data to the device control, connected to the control or monitoring system.

8.5   COMPONENTS USED BY IoT

Different types of sensors and components are used to control the overall working progress of the monitoring systems in IoT [21]. There are some components required for monitoring progress. The components are

- RFID
- Arduino/Raspberry Pi
- Relay
- Sensors
- Bluetooth/Wi-Fi
- Data Transmission/Data Analysis

8.6   RASPBERRY PI/ARDUINO

The overall circuit process is controlled by the Raspberry or Arduino Uno 3 board. It can be connected with the device, which needs to sense the data and overall monitoring process. These boards work as a centralized control unit. Each node is connected within the range of the circumstance [22, 23]. The entire control unit process is based on any one of Arduino or Raspberry Pi board.

Most of the devices use Raspberry Pi because of the low cost and size of the component. The main advantage of Raspberry is the very small-sized card that acts

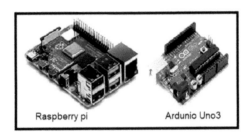

FIGURE 8.2   Processing unit Raspberry Pi and Ardunio Uno3.

as a computer that may be plugged into a computer's monitor or any other component like TV, etc., and it uses a QWERTY keyboard and a mouse [27, 28]. It can
develop small devices that enable people of all ages to easily explore new computing
devices and easily learn programming ethics like python scratch and MATLAB.
It's capable of making the application compatible to the desktop to do so, from
surfing the Internet and making the decision based on the data analysis report, playing
the high-definition audio and video, to making powerpoint presentations, spreadsheets,
data-driven applications, word document, word-processing, playing games, etc. [31].

Next, even kids worldwide use the Raspberry Pi for program learning and easily
understand manipulating the computer [32, 33]. And its main advantage is the ability
of interaction within the outside world, and it is used in the wider range of array for
the all digitalized component projects, from video, music, machines and patient
detectors to all like weather monitoring and infrared cameras and tweeting, etc. [38]
(Figure 8.2).

The Arduino Uno 3 is a microcontroller based on Atmel's ATmega328. It has 14
pins as digital input/output pins. Since it may have 6 as digital input and 6 as analog
inputs, this board can connect with USB and have external power intrusion[43]. It
can simply connect to the computer with the help of a USB driver cable. Even it is
capable of acceptance from macOS, Linux,etc.

8.7   RFID TAG

The Radio Frequency Identification Tag is one of the tracking systems. It is used in
smart barcodes to identify the item which has been attached to the security systems.
The RFID tags utilize radio frequency technology [51] (Figure 8.3).

It consists of an integrated circuit with a transistor antenna. It may help to transmit
the data to the RFID tag reader component. Then the reader converts that reader data
into the form of radio signal waves. The collected information is transferred to a data
server with the help of a communication interface for data analysis and computerized
decision-making progress [73].

8.8   CONDITION RELAY

A relay is one of the components which are operated manually or electrically by
automatic devices. It has two types of device systems: (1) a control system and (2) a
controlled system. It is almost used for all other electronic circuit-based devices.

FIGURE 8.3   RFID tag implemented components.

FIGURE 8.4   Different types of relay.

It can handle the device with a high current within the low-consumption power display. It can find the lower inertia moving objects and stability, long-term reliability with the small volume. It is used widely for the power consumption system and all automation technology, sports, remote control devices, surveillance and communication, and devices of electro-mechanics and all power-oriented electronics [47] (Figure 8.4).

Generally, a relay contains the induction part to monitor the power system. It can exhibit input data like voltage passed, current usage, power consumption, resistance, speed control, wide temperature, data frequency signal, pressure, light, etc. And especially, it also contains the actuator module, which can easily find the pressure cage and data validation, which can energize or de-energize the connection made by the controlled circuit. An intermediate component will follow the input and output coupling and cohesion for the current input and current accuracy [52].

The relay has a major role in controlling the wireless device in the relay (ESP8266) to enforce wireless communication and to make them effective with the help of an IR sensor. By object detection, the relay will activate the GPRS to read the data enclosed with the corresponding object.

8.9   BLUETOOTH/WI-FI

The GPRS tracking system and all other techniques lead to wireless communication. These modems can make this. This is a kind of Tag that may be fixed with IoT-based devices to make the communication channel between nodes. The RFID tag manages

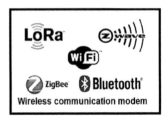

FIGURE 8.5   Wireless modems.

the data and the transmitter transfers the sensed data and tracked details to the cloud. All these are controlled and managed by Arduino Uno 3/Raspberry Pi (Figure 8.5).

8.10   DATA SERVER

Each process of the device generates the data. It is kept on to permanent memory through the communication channel and RFID Tag. The data server collects data transmission and data analysis as even the security alert communication messages.

The data storage in IoT is cloud storage. To keep all the data in a database to manage the inefficient data way. The information is transferred to a cloud memory detected by the sensor and the relay. These databases are kept in the form of data as a digital signal with the help of an RFID tag.

Initially, the PIR sensor will sense the data about the object that has been detected. Then it will activate the other sensor and also an RFID tag to detect the devices. Then it will keep on reading the term value and compare it with all existing data and condition factors. Based upon the outcome, it may be responsible for the decision-making process. Then finally, the data will be transferred to the cloud at the same time, it will find the statistical form data for every individual record [67].

8.11   DATA ANALYSIS: THINGSPEAK

It is an open-source application. This platform may provide the data analysis and visualization process in an efficient manner. It also provides the service to allow visualization, analysis and specified live data streaming onto the cloud data storage server. Its instant information about the visualization data representation helps in frequent access and monitoring progress. It can execute even a MATLAB code onto the server and immediately evaluate and process the statistical data analysis [69]. It is regularly used for prototyping.

8.11.1   Features of ThingSpeak

- It is easily configured to transmit data to cloud memory.
- It has the ability to visualize the actual time of sensor activity.
- It collects data on demand from different resources.
- It builds IoT-based systems and prototyping without developing any special software and servers.

8.12   SENSORS

A device that responds to a physical interaction with the object, such as temperature, heat, sound waves, pressure cage volume, light-emitting, magnetism, or any other motions, transmits a resulting data report in measurement or operating control.

Sensors are sophisticated devices that can be frequently used to detect the input data as suspected detail and respond to an electrical or optical signal. The sensors convert the data into signals for the data transmission [73] (Figure 8.6).

Sensors are widely applied everywhere, like radar guns, drone cameras, lens activators, rocket launches, bounces, microwaves, cars, elevators, escalators, etc. An alarming burglar system may use a sensor to find the power failure or water insufficient in the washing machine, or the ultrasonic sensor will find the moving object using the sound waves. Likewise, different sensors are available; they may sense the object's actuator pressure and temperature, health factors like blood pressure, heartbeat rate, etc. Stud finders are used to find the location of woodenXstuds by the carpenters under a wall. It may employ magnets or any other like radar.

8.13   TYPES OF SENSORS

There are various sensors used in real-time applications [85]. Based on the requirement of the monitoring systems, many of the sensors take part in the manipulation.

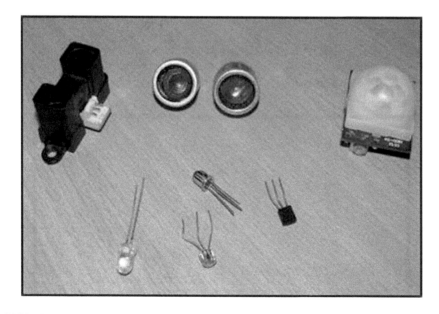

FIGURE 8.6   Commonly used sensors.

8.13.1 PIR Sensor

The Passive Infrared Sensor is a major component for detecting moving objects within a wide range of surfaces. It finds the object or human or any object coming near to the tunnel. At that time of object detection, the RFID Tag will activate the object sensor while activating the GPRS sensor and all other related sensor nodes [86] (Figure 8.7).

For this purpose, the IRA-S210ST01 PIR sensor has been affordable, which contains a vast surface to detect the object and consumes lesser memory in Arduino.

8.13.2 Temperature Sensor

This sensor collects information about the temperature of the corresponding object on the surface. Then it converts into the form of device understandable formats. The best implementation of the temperature sensor is contactless sensing. The temperature sensor considers the outspace measurement as a source element. The temperature sensors can be classified into two types (Figure 8.8):

Contact Sensors – This type of sensor needs the physical interaction of the object or media that is being sensed.
Non-contact Sensors – This type of sensor does not require any physical interaction with the object or any other media which is being sensed.

8.13.3 IR Sensor

The basic idea is to make use of IR LEDs to send infrared waves to the object. Another IR diode of the same type is used to detect the reflected wave from the object (Figure 8.9).

This device finds the infrared radiation to the particular environmental surface area. And it may penetrate the IR waves to the particular object for the kind of manipulation work. This IR sensor emits the IR rays for human eye optical capacity [35].

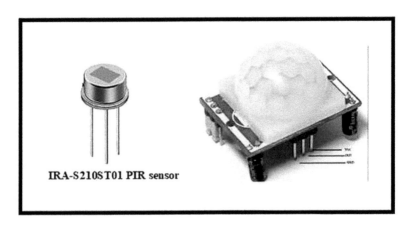

IRA-S210ST01 PIR sensor

FIGURE 8.7 PIR sensor.

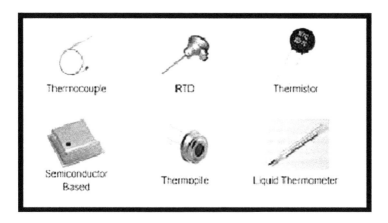

FIGURE 8.8   An image showing thermocouple different temperature sensor.

FIGURE 8.9   IR Sensor.

8.13.4   UV Sensor

The ultraviolet sensor is used to find the object's motivation using the UV rays. This kind of technology is applied in baby scanning and the like quotable screening and related activities.

8.13.5   Touch Sensor

A touch sensor acts as an individual variable resistor for the particular device as the location finder could find the touching interaction. The touch sensor is shown in Figure 8.10.

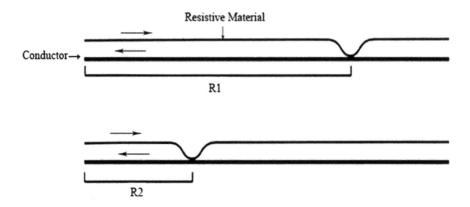

FIGURE 8.10   Touch sensor resistor.

8.14   IoT-BASED MONITORING SYSTEM APPLICATIONS

8.14.1   HEALTH MONITORING SYSTEM

Good medical maintenance is to provide better treatment at the required time. Internet makes it easier to monitor the loved ones for the kind of medical care they need. This system helps to monitor the patient's health progress like a regular temperature check, ECG, heartbeat rate, EMG, motion sensor, or any other bad condition cases symptoms. If there are any particular abrupt changes in any patient's heart rate or body temperature, that will alert the attached system about the patient using IoT [37] (Figure 8.11).

After the patient's major operations, the patient is advised to stay at home and their medical condition is monitored by special caretakers or any family members only. Many people are working. They can't spend time to monitor the medical cases or any old age people because of their necessity to work for the environment. To overcome these kinds of issues the patient's health is monitored from anywhere using IoT. This technology with the micro-controller and also the Wi-Fi module helps the user monitor their loved ones. The integral development of young children is fundamental to human progress. It finds the disorders, geo-location, harassment-related activity, medical emergencies, and their health and safety. It helps to manage them easily and in a proper way of guidance [43].

8.14.2   TEMPERATURE MONITORING SYSTEM

It can easily monitor the human temperature of those who crossed the tunnel and also keep their data securely. This reduces the need for human safety to stay at a distance from the people affected by that virus. To avoid the number of human resources involved in monitoring at any kind of public sector, it senses their mobile number with the help of a GPS tracking system. In personally guiding the people and what kind of necessary action they must take and symptoms, all other related information (Figure 8.12).

With the help of this system, we can avoid the manual checking of the human temperature, and reduce the cause for the spread of virus. Especially, we can avoid the community spread over the working environment and could manage the details of

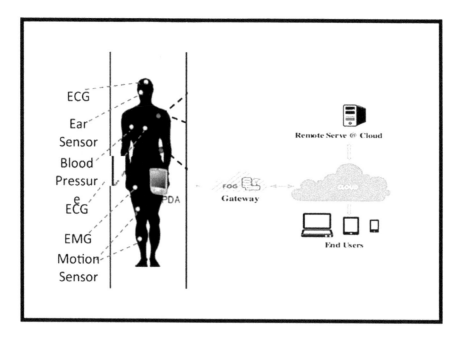

FIGURE 8.11   General architecture for health monitoring system.

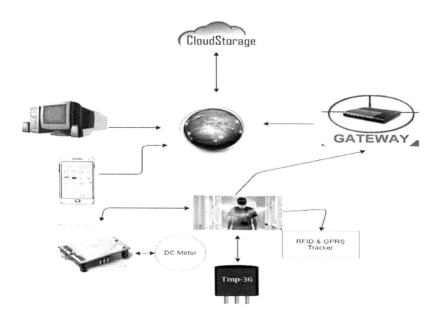

FIGURE 8.12   Temperature monitoring system architecture.

every employee securely. At the same time, it guides the affected person without knowing others. It secretly handles them and gives better suggestions for the necessary action they must take for critical situations. In the future, it can be implemented with all thumbs or a face-recognizing attendance register. And it keeps track of the data in cloud memory [45].

8.14.3 SMART IRRIGATION SYSTEM FOR AGRICULTURE

This smart irrigation system for agriculture is used to find the moisturizer level of the solids and may help them maintain easily without the need for any physical visit. Likewise, many applications are developed for agriculture. For example, to check the farming plant level for avoiding unnecessary irrigation. Then even to avoid the wild animals from entering the farming land. They have done a major part of the cultivation now. It improved technology to monitor the cultivation of land, and even it may give the alarming system better than this and may take necessary actions based on the land situation. The resultant data should be kept on to the cloud for any data analysis [66] (Figure 8.13).

A cultivation land monitoring system will provide a better environment to keep the cultivation very formal. It also keeps the analysis report for further action to be taken for the issues that may arise from any critical region.

8.15 MACHINE MONITORING SYSTEM

The core companies are also using this IoT-based technology to keep the 24/7 miniaturization of the production and all other related issues. This sensed data would help them avoid the accident and any other issues. These systems are providing high productivity, self-decision-making and reproductivity status; self-configurable situation may be handled smoothly. And also the production of high-quality products in all manufacturing progress lines [36] (Figure 8.14).

The S-CMS is to provide the environment in the industries to avoid the fault conditions. It automatically detects machine failures. It may find the root causes of the failure; it will take the self-configurable and self decision-making will help improve the plant's productivity.

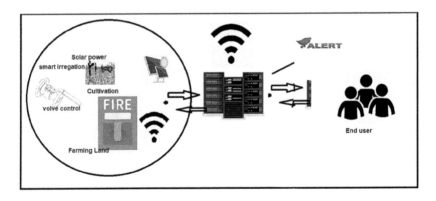

FIGURE 8.13 Architecture for smart agriculture.

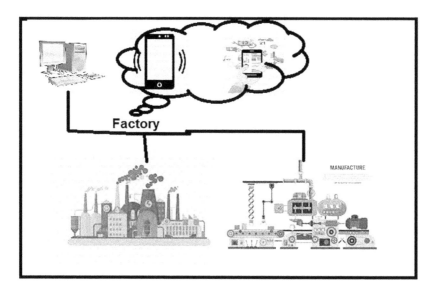

FIGURE 8.14 Smart factory monitoring system architecture.

8.16 MARINE RADIOACTIVITY MONITORING SYSTEM

The marine radioactive data are incomplete; it will cause spectrum errors when the ship tries to contact anyone for emergency cases. By that time, the packets sending as a single unit will not be possible. It sends the packet at a low spectrum range and even avoid the intruder's known activity to keep into the smaller packet in the cipher text. It continuously monitors marine activity like the marine environment, any other suspected activities, and radar-based activities (Figure 8.15).

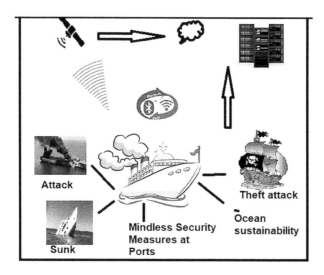

FIGURE 8.15 Architecture of marine radioactivity.

The system can monitor marine radioactivity within an infinite time duration. It uses low power with automatic long-term testing—the early warning of marine nuclear safety and emergency response [13].

8.17   VEHICLE TRACKING AND FUEL MONITORING SYSTEM

To track a particular vehicle will utilizing GPS and then GSM. Accident detection alert system, soldier tracking system and fuel monitoring system [53]. It is widely used for all the environment like school transport, cabs, rental vehicle services, mobile bike, police patrol and rescue vehicle in mishaps. The vehicle tracking system will find the vehicle's location and even find the vehicle number without being known by the suspected person. The flame and water-level monitoring helps to monitor the flame accident and the water level. It is a model of assistance of GPS and beneficiaries of the GSM modem itself (Figure 8.16).

8.18   ANDROID-BASED MONITORING SYSTEM

The monitoring system is moving on to the next level to monitor from the mobile-based application. It helps control the IoT-based devices from handheld devices and keeps that data in cloud memory. Especially to track the data with the geolocation-based monitoring system, this kind of location tracking system is accessed from mobile devices very efficiently.

This health monitoring system is useful for 24/7 monitoring patient health like blood pressure, heartbeat rate, etc. But it doesn't require the patient to visit the hospital and reduce the unwanted doctor's visit [35].

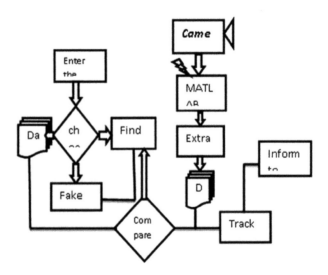

FIGURE 8.16   Vehicle tracking and fuel monitoring system.

8.19   APPLIED ALGORITHM

8.19.1   PERFORMANCE OF SYMMETRIC KEY

The symmetric key algorithms have a smaller amount of key size, smaller memory usage, and less time for computations. This may be perfectly suitable for communication between a smaller number of users. The medical-oriented data is accessed by few users only. The block cipher is 64-128 bits in size. The RC6 and blowfish symmetric key encryption techniques are used.

8.19.2   BLOWFISH ALGORITHM

The key length is 32 bit to 448 bits. It has 16 Feistel cipher routines. It's more secure than an s-box, in every round (r), consisting of four steps. The key will be separated into 8 bit each. Each bit can be considered as a left and right bit. The left bit will be XORed and taken into the right of the next round of the Feistel. The F-function's output is XORed with the right half (R) of the data. Then the left half (L) key will be a right half(r), and the right half (R) is being left half (L), both can be swapped [23].

8.19.3   CIPHER-ATTRIBUTE-BASED ENCRYPTION

The private key of each user is associated with the key permutations. It consists of four steps. Each step may consist of different techniques of individual key conclusions [36].

The first step set the Primary Key (PK) and Master Key (MK) to define the security key encryptions.

A second step will do the encryption process; the original message has been converted into cipher text based on the key PK and MK.

The third step to KeyGen is nothing but key encryption techniques.

At last, the step is to decrypt the message using the decrypted key and cipher text, and it also does almost all 16 rounds of structure and then gets the original message text as M.

8.20   CONCLUSION

IoT-based monitoring systems are used to avoid the manual checking monitoring process and their detail without the physical interference of the device, to avoid the unwanted people who may enter on the public administrations and especially to avoid the community spread of disease to the working environment and to manage the details of every employee securely. At the same time, it guides the affected person without knowing others. It secretly handles them and gives better suggestions for the necessary action they must take for critical situations and it keeps track of the data in cloud memory. In the future, it can able to implement with all thumbs or a face-recognizing attendance register.

REFERENCES

1. Joes K. Reena & R. Parameshwari, "*A Smart Health Care Monitor System in IoT Based Human Activities of Daily Living: A Review*", *2019 International Conference on Machine Learning (COMITCon)*.
2. Lusi Susanti & Dicky Fatrias, "*A Configuration System for Real-Time Monitoring and Controlling Electricity Consumption Behavior*", *IET*, 2019.
3. B. Praveen Kumar, K. Mahendrakan, "*Prevention of Accident due to Drowsy by Using Eye Blink*", *IJIRSET*, vol. 3, issue 5, May 2014.
4. Vikas Desai, "Design and Implementation of GSM and GPS Based Vehicle Accident Detection System", *IJIT*, Vol. 1, Issue 3, 2013.
5. S. Vigneshwaran, B. Nithya, "*Design of Bus Tracking & Fuel Monotoringsystem*", *ICACCS*, 2020.
6. K. V. S. Baba, S. R. Narasimhan, "Synchrophasor Based Real Time Monitoring of Grid Events in Indian Power System", 978-1-4673-8848-1/16/, 2016, IEEE.
7. D. Shiva Rama Krishnan, Subhash Chand Gupta, "*An IoT Based Patient Health Monitoring System*", *2018 International Conference on Advances in Computing and Communication Engineering (ICACCE-2018)*, Paris, France, 22–23 June 2018.
8. M. Omoogun, V. Ramsurrun, S. Guness, P. Seeam, X. Bellekens and A. Seeam, "*Critical Patient eHealth Monitoring System Using Wearable Sensors*", *1st International Conference on Next Generation Computing Applications (NextComp)*, Mauritius, 2017, pp. 169–174. doi: 10.1109/NEXTCOMP.2017.8016194.
9. P. Kakria, N.K. Tripathi, P. Kitipawang, "A Real-Time Health Monitoring System for Remote Cardiac Patients Using Smartphone and Wearable Sensors", *International Journal of Telemedicine and Applications*, 2015, 373474. doi:10.1155/2015/373474.
10. https://en.wikipedia.org/wiki/Passive_inIOT_infrared_sensor.
11. "Heealthcare Survey Including the Networks and Architecture", http://ieeexplore.ieeesorg/document/7917920/.
12. "*Report on Power System Oscillations Experienced in Indian Grid on 9th, 10th, 11th and 12th August 2014*", *POSOCO*, India.
13. "*Operational Feedback on Transmission Constraints*", *POSOCO*, April 2016.
14. "*Report on Low Frequency Oscillation in Indian Power System*", *POSOCO*, India, March 2016.
15. B. Massot, C. Gehin, R. Nocua, A. Dittmar, E. McAdams, "*A Wearable Low Power Health Monitoring Instrumentation Based on Programming System on Chip TM*", *Annual International Conference of the IEEE Engineering in Medicine and Biology Society*, 2009.
16. S. Devendra, K. Verma, P.K. Barhai, "*Design and Development of WINGSNET (Wireless Intelligent GPS-Based Sensor Network) System for Monitoring Air Pollution and Radiation Based on Wi-Fi & WiMAX Communication Network*", *2014 IEEE 11th International Conference on Mobile Ad HocandSensorSystems*.
17. H. Zhang, F. Zhang, Y. Wang, G. Zhang, "*Wireless Sensor Network Based Anti-Theft System of Monitoring on Petroleum Pipeline*", *2011 Second International Conference on Mechanic Automation and Control Engineering*.
18. R. Carlos, S. Coyle, B. Corcoran, D. Diamond, Ward Tomas, McCoy Aaron, Florin Stroiescu, Kieran Daly, "Web-Based Sensor Streaming Wearable for Respiratory Monitoring Applications", *Sensors*, 2011, IEEE.
19. N. Barie, A. Skrypnik, A. Voigt, M. Rapp, J. Marcoll, "*Work Place Monitoring Using a High Sensitive Surface Acoustic Wave Based Sensor System*", *TRANSDUCERS 2007—2007 International Solid-State Sensors, Actuators and Microsystems Conference*.

20. T. Muthamilselvan et al., "We Are IntechOpen the World's Leading Publisher of Open Access Books Built by Scientists for Scientists TOP 1% Control of a Proportional Hydraulic System", *Intech Open*, vol. 54, pp. 713–727, June 2016.
21. D. Perakovi and M. Peri, *"Advances in Design Simulation and Manufacturing*, Springer International Publishing, vol. 1, 2019.
22. B. Okonokhua, B. Ikhajiagbe, G. Anoliefo and T. Emede, "The Effects of Spent Engine Oil on Soil Properties and Growth of Maize (*Zea mays* L.)", *J. Appl. Sci. Environ. Manag.*, vol. 11, no. 3, 2010.
23. J. A. Heredia-Cancino, M. Ramezani and M. E. Alvarez-Ramos, "Effect of Degradation on Tribological Performance of Engine Lubricants at Elevated Temperatures", *Tribol. Int.*, vol. 124, pp. 230–237, 2018.
24. J. Ma, Z. Zong, F. Guo, Y. Fei and N. Wu, "Thermal Degradation of Aviation Synthetic Lubricating Base Oil", *Pet. Chem.*, vol. 58, no. 3, pp. 250–257, 2018.
25. S. Zzeyani, M. Mikou and J. Naja, "Physicochemical Characterization of the Synthetic Lubricating Oils Degradation under the Effect of Vehicle Engine Operation", *Eurasian J. Anal. Chem.*, vol. 13, no. 4, 2018.
26. S. Feng, B. Fan, J. Mao, Y. Xie and Y. Che, "An Oil Monitoring Method of Wear Evaluation for Engine Hot Tests", *Int. J. Adv. Manuf. Technol.*, vol. 94, no. 9–12, pp. 3199–3207, 2016.
27. K. Azevedo and D. B. Olsen, *"Engine Oil Degradation Analysis of Construction Equipment in Latin America"*, J. Qual. Maint. Eng., 2019.
28. T. Holland, A. M. Abdul-Munaim, D. G. Watson and P. Sivakumar, "Influence of Sample Mixing Techniques on Engine Oil Contamination Analysis by Infrared Spectroscopy", *Lubricants*, vol. 7, no. 1, pp. 4, 2019.
29. S. M. Azzam et al., "Characterization of Essential Oils from Myrtaceae Species Using ATR-IR Vibrational Spectroscopy Coupled to Chemometrics", *Ind. Crops Prod.*, vol. 124, pp. 870–877, February 2018.
30. L. Hong and K. Sengupta, *"Fully Integrated Optical Spectrometer with 500-to-830 nm Range in 65nm CMOS"*, *Dig. Tech. Pap. - IEEE Int. Solid-State Circuits Conf.*, vol. 60, pp. 462–463, 2017.
31. "Sunrom", *Light Dependent Resistor - LDR*, p. 4, 2008.
32. P. Onorato, L. M. Gratton, M. Polesello, A. Salmoiraghi and S. Oss, "The Beer Lambert Law Measurement Made Easy", *Phys. Educ.*, vol. 53, no. 3, 2018.
33. R. Khan, S. U. Khan, R. Zaheer and S. Khan, *"Future Internet: The Internet of Things Architecture Possible Applications and Key Challenges"*, *Proc. - 10th Int. Conf. Front. Inf. Technol. FIT 2012*, pp. 257–260, 2012.
34. C. Z. Diego and A. P. Giménez, *Edible and Medicinal Mushrooms: Technology and Applications*, Chennai: John Wiley & Sons Ltd, 2017.
35. M. Singh, B. Vijay, S. Kamal and G. Wakchaure, *Mushrooms: Cultivation Marketing and Consumption*, Himachal Pradesh: Directorate of Mushroom Research, 2011.
36. T. Iwade and V. Mizuno, "Cultivation of Kawariharatake (Agaricusblazei Murill)", *Food Reviews International*, vol. 13, no. 3, pp. 383–390, 1997.
37. T. Kaewwiset and P. Yodkhad, *"Automatic Temperature and Humidity Control System by Using Fuzzy Logic Algorithm for Mushroom Nursery"*, *International Conference on Digital Arts Media and Technology (ICDAMT)*, pp. 396–399, 2017.
38. P. Shiombing, T. P. Asturin, Herriyance and D. Sitompul, *"Microcontroller Based Automatic Temperature Control for Oyster Mushroom Plants"*, *Journal of Physics: Conference Series*, vol. 978, p. 012031, 2018.
39. F. Mohammed, A. Azmi, Z. Zakaria, M. F. N. Tajuddin, Z. M. Isa and S. A. Azmi, "IoT Based Monitoring and Environment Control System for Indoor Cultivation of Oyster Mushroom", *Journal of Physics: Conference Series*, vol. 1019, pp. 012053, 2018.

40. R. Y. Adhitya et al., *"Comparison Methods of Fuzzy Logic Control and Feed Forward Neural Network in Automatic Operating Temperature and Humidity Control System (Oyster Mushroom Farm House) Using Microcontroller"*, International Symposium on Electronics and Smart Devices (ISESD), pp. 168–173, 2016.

41. G. Zervakis, A. Philippoussis, S. Ioannidou and P. Diamantopoulou, "Mycelium Growth Kinetics and Optimal Temperature Conditions for the Cultivation of Edible Mushroom Species on Lignocellulosic Substrates", *Folia Microbiologica*, vol. 46, no. 3, pp. 231–234, 2001.

42. P. Pal et al., "IoT Based Air Pollution Monitoring System Using Arduino", *IRJET Journal*, vol. 4, no. 10, October 2017.

43. C. V. Saikumar, M. Reji and P.C. Kishoreraja, "IOT Based Air Quality MonitoringSystem", *International Journal of Pure and Applied Mathematics*, vol. 117, no. 9, pp. 53–57, 2017.

44. "The Sensor Web: A New InstrumentConcept", *SPIE Symposium on Integrated Optics*, January 2003.

45. P. Völgyesi, A. Nádas, X. Koutsoukos and Á. Lédeczi, *"AirQuality Monitoring with SensorMap"*, *Proceedings of the 7thinternational Conference on information Processing in Sensor Networks*, pp. 529–530, 2008.

46. Sonal A. Mishra, Dhanashree S. Tijare and G. M. Asutkar, "Design of Energy Aware Air Pollution Monitoring System Using WSN", *International Journal of Advances in Engineering & Technology*, vol. 1, no. 2, pp. 107–116, May 2011.

47. R. A. Roseline, M. Devapriya and P. Sumathi, "Pollution Monitoring Using Sensors and Wireless Sensor Networks: A Survey", *International Journal of Application or Innovation in Engineering & Management (IJAIEM)*, vol. 2, no. 7, pp. 119–124, July 2013.

48. Nikheel A. Chourasia and Surekha P. Washimkar, *"ZigBeeB ased Wireless Air Pollution Monitoring"*, *International Conference on Computing and Control Engineering (ICCCE 2012)*, 12 and 13 April 2012.

49. Tsow, E. Forzani, A. Rai, R. Wang, R. Tsui, S. Mastroianni, et al., "A Wearable and Wireless Sensor System for Real-Time Monitoring of Toxic Environmental Volatile Organic Compounds", *IEEE Sensors J.*, vol. 9, pp. 1734–1740, Dec. 2009.

50. Y. J. Jung, Y. K. Lee, D. G. Lee, K. H. Ryu and S. Nittel, "Air Pollution Monitoring System Based on Geo Sensor Network", *Proc. IEEE Int. Geo science Remote Sensing Symp*, vol. 3, pp. 1370–1373, 2008.

51. W. Chung and C. H. Yang, "Remote Monitoring System with Wireless Sensors Module for Room Environment", *Sens. Actuators B*, vol. 113, no. 1, pp. 35–42, 2009.

52. A. Whitmore, A. Agarwal and L. Da Xu, "The Internet of Things—A Survey of Topics Andtrends", *Information Systems Frontiers*, vol. 17, pp. 261–274, 2015.

53. S. Li, L. D. Xu and S. Zhao, "The Internet of Things: A Survey", *Information Systems Frontiers*, vol. 17, pp. 243–259, 2015.

54. Q. Sun, J. Liu, S. Li, C. Fan and J. Sun, "Internet of Things: Summarize on Concepts Architecture and Key Technology Problem", *Journal of Beijing University of Posts Telecommunications*, vol. 33, pp. 1–9, 2010.

55. IEEE 1451 Smart Transducer Interface Standards, [online] Available at: www.nist.gov/el/isd/ieee/IEEE1451.cfm.

56. Q. Chi, H. Yan, C. Zhang, Z. Pang and L. D. Xu, "A Reconfigurable Smart Sensor Interface for Industrial WSN in IoT Environment", *IEEE Transactions on Industrial Informatics*, vol. 10, pp. 1417–1425, 2014.

57. Kumar, I. P. Singh and S. K. Sud, "Energy Efficient and Low-Cost Indoor Environment Monitoring System Based on the IEEE 1451 Standard", *IEEE Sensors Journal*, vol. 11, pp. 2598–2610, 2011.

58. A. Kumar, V. Srivastava, M. K. Singh and G. P. Hancke, "Current Status of the IEEE 1451 Standard- Based Sensor Applications", *IEEE Sensors Journal*, vol. 15, pp. 2505–2513, 2015.

59. [online] Available at: http://china.xilinx.com/support/documentation/user_guides/ug585-Zynq-7000-TRM.pdf.

60. L. Hu, H. Li, X. Xu and J. Li, "*An Intelligent Vehicle Monitoring System Based on Internet of Things*", *7th International Conference on Computational Intelligence and Security CIS 2011*, pp. 231–233, 2011.

61. P. Pyykonen, J. Laitinen, J. Viitanen, P. Eloranta and T. Korhonen, "*IoT for Intelligent Traffic System*", *IEEE 9th International Conference on Intelligent Computer Communication and Processing ICCP 2013*, pp. 175–179, 2013.

62. M. Al-Kuwari, A. Ramadan, Y. Ismael, L. Al-Sughair, A. Gastli and M. Benammar, "*Smart-Home Automation Using IoT-Based Sensing and Monitoring Platform*", *2018 IEEE 12th International Conference on Compatibility Power Electronics and Power Engineering (CPE-POWERENG 2018)*, pp. 1–6, 2018.

63. Xie Lu, *Supervisory Control and Data Acquisition System Design for CO2 Enhanced Oil Recovery*, May 2014.

64. R. Shete and S. Agrawal, "*IoT Based Urban Climate Monitoring Using Raspberry Pi*", *2016 International Conference on Communication and Signal Processing (ICCSP) Melmaruvathur*, pp. 2008–2012, 2016.

65. M. Lekić and G. Gardašević, "*IoT Sensor Integration to Node-RED Platform*", *2018 17th International Symposium INFOTEH-JAHORINA (INFOTEH) East Sarajevo*, pp. 1–5, 2018.

66. A. Rajalakshmi and H. Shahnasser, "*Internet of Things Using Node-Red and Alexa*", *2017 17th International Symposium on Communications and Information Technologies (ISCIT)*, pp. 1–4, 2017.

67. Li Wang and K. Liu, "*Implementation of a Web-Based Real-Time Monitoring and Control System for a Hybrid Wind-PV-Battery Renewable Energy System*", *2007 International Conference on Intelligent Systems Applications to Power Systems*, pp. 1–6, 2007.

68. N. P. Kumar and R. K. Jatoth, "*Development of Cloud Based Light Intensity Monitoring System Using Raspberry Pi*", *2015 International Conference on Industrial Instrumentation and Control (ICIC)*, pp. 1356–1361, 2015.

69. V. Sandeep, K. L. Gopal, S. Naveen, A. Amudhan and L. S. Kumar, "*Globally Accessible Machine Automation Using Raspberry Pi Based on Internet of Things*", *2015 International Conference on Advances in Computing Communications and Informatics (ICACCI)*, pp. 1144–1147, 2015.

70. S. Chanthakit and C. Rattanapoka, "*MQTT Based Air Quality Monitoring System Using Node MCU and Node-RED*", *2018 Seventh ICT International Student Project Conference (ICT-ISPC)*, pp. 1–5, 2018.

71. J. C. B. Lopez and H. M. Villaruz, "*Low-Cost Weather Monitoring System with Online Logging and Data Visualization*", *2015 International Conference on Humanoid Nanotechnology Information Technology Communication and Control Environment and Management (HNICEM)*, pp. 1–6, 2015.

72. January 2019, [online] Available at: https://emoncms.org/.

73. January 2019, [online] Available at: https://www.abbey.co.nz/papers.html.

74. S. Patil, M. Vijayalashmi and R. Tapaskar, "Solar Energy Monitoring System Using IoT", *Indian J. Sci. Res*, vol. 15, pp. 149–155, 2017.

75. W. Energy, "*Executicve Summary of the World Energy Investment 2018*", Int. Energy Agency, 2018.

76. A. Woyte, M. Richte, D. Moser, S. Mau, N. Reich and U. Jahn, *"Monitoring of Photovoltaic Systems"*, *Proc. 28th European Photovoltaic Solar Energy Conference*, pp. 3686–3694, 2013.

77. F. Shariff, N. Abd Rahim and W. Hew, "Zigbee-Based Data Acquisition System for Online Monitoring of Grid-Connected Photovoltaic System", *Expert Systems with Applications*, vol. 42, no. 3, pp. 1730–1742, 2015.

78. S. Madeti and S. Singh, "Monitoring System for Photovoltaic Plant: A Review", *Renewable and Sustainable Energy Reviews*, vol. 67, pp. 1180–1207, 2017.

79. A. Hamied, A. Mellit and M.A. Zoulid, *"IoT-Based Experimental Prototype for Monitoring of Photovoltaic Arrays"*, *IEEE 1st International Conference on Applied Smart Systems ICASS18*, 24–25 November 2018.

80. Q. Ou, Y. Zhen, X. Li, Y. Zhang and L. Zeng, *"Application of Internet of Things in Smart Grid Power Transmission"*, *IEEE Third FTRA International Conference on Mobile Ubiquitous and Intelligent Computing*, 2012.

81. A. Mellit and S. Kalogirou, "MPPT-Based Artificial Intelligence Techniques for Photovoltaic Systems and Its Implementation into FPGA Chips: Review of Current Status and Future Perspectives", *Energy*, vol. 70, pp. 1–21, 2014.

82. L. Bouselham, B. Hajji, A. Mellit, A. Rabhi and A. Mazari, *"A Reconfigurable PV Architecture Based on New Irradiance Equalization Algorithm"*, *International Conference on Electronic Engineering and Renewable Energy*, pp. 470–477, 2018.

83. A. Mellit, H. Rezzouk, A. Messai and B. Medjahed, "FPGA-Based Real Time Implementation of MPPT Controller for Photovoltaic Systems", *Renewable Energy*, vol. 36, pp. 1652–1661, 2011.

84. S. Motahhir, A. Chalh, A. El Ghzizal and A. Derouich, "Development of a Low-Cost PV System Using an Improved INC Algorithm and a PV Panel Proteus Model", *Journal of Cleaner Production*, vol. 204, pp. 355–365, 2018.

85. S. Belhimer, M. Haddadi and A. Mellit, "A Novel Hybrid Boost Converter with Extended Duty Cycles Range for Tracking the Maximum Power Point in Photovoltaic System Applications", *International Journal of Hydrogen Energy*, vol. 43, pp. 6887–6898, 2018.

9 A Progressive Method to Monitor Power Using IoT

S. Suriya, Agusthiyar Ramu, and J. Shyamala Devi
SRM IST Ramapuram Campus, Chennai, India

CONTENTS

9.1 INTRODUCTION

Power Monitoring System plays a vital role in every household. In the modern world, the new arrival of electronic gadgets is increasing. The power consumed by each device increases with the use of each new gadget, which leads to a high electricity bill.

DOI: 10.1201/9781003156123-9

To avoid this kind of situation, a novel method to monitor power is very useful and offers a wonderful way which predicts how much power is used daily, weekly, or monthly. Energy consumption, accompanied by the real-time monitoring is achieved by the smart system and operates on the IoT concept. Keeping track of the power consumption of each gadget at regular intervals gives a clear idea to visualize the power consumed by each device. Home automation accompanied by a power monitoring system gives good clarity on the daily usage of electricity. Daily usage of power consumption may be plotted on the Graphical User Interface (GUI). GUI helps us visualize the abnormal usage of any user's grade and can easily detect the theft of power.

9.2 LITERATURE STUDY

Power consumption and monitoring is the need of the hour. The electric power monitoring systems monitor and control electric power equipment, including the power receiving and transforming facilities installed in buildings, plants, and in the transportation sector. Moreover, power consumption can conserve energy by utilizing the electric power demand monitoring function and energy visualization function [1]. Power consumption is measured according to a set cycle to prevent power consumption from exceeding contracted power, producing an alarm when power consumption is predicted to exceed contracted power [2]. Several research findings regarding power monitoring and different approaches are proposed to precisely predict energy usage and the individual can plan accordingly [3].

Hugo Tavares et al. [2] in their paper mention about electric power consumption from remote using non-intrusive smart power. The objective is monitoring energy in real time with a minimal cost sensor. By using the MQTT protocol, the module can be reused in any project. The noninvasive sensors and measurement of power factor work in the emonLib library. Wi-Fi-enabled technology helps communication between the meter and outside environment, which allows observing the data through several web servers and excel sheets in real time. It also uses low-cost current sensors [4].

9.2.1 System Design to Detect Load Automatically

The system design workflow is based on automatically detecting load by using software on hardware design. The hardware design for the load detection system is automated using software, and the workflow comprises four steps:

9.2.1.1 Collection of Data on Power Consumption

The collection of data for power consumption is different for start, stop, and switching. Data collection using trivial methods is categorized and compared with the own-data novel acquisition method from the gadgets. The data collected by the sensor is stored in the grid load database and also updated at regular intervals.

9.2.1.2 Extracting Data of Power Load

A lot of methods to extract data are proposed by many authors, and the data can be extracted using any trivial method possible. After the extraction of data, another extraction for power load with different features is done. These all things happen for achieving accuracy as it may affect the monitoring directly.

9.2.1.3   Processing of Power Load Data

The next step after extracting data for usage of power is the processing of data. After extraction, the data extracted must be processed to convert detailed data into digital code and edited load code. After that, the filtration needs to happen for uniformly collected data on abnormality clearance and unwanted data. The real-time database stores the valid data to be processed.

9.2.1.4   Tracking of Data

The real-time process contains data monitoring, then the corresponding results are achieved by the real-time process. Data should be tracked and the values are different with graphical representation to be displayed. This promotes us to analyze better, and monitoring of the result can be done.

Jisoo Park et al. [4] propose a prediction model based on deep learning, where the power bill is predicted considering climatic conditions. A gated recurrent unit-based model that integrates weather and date and predicts power usage for each household. Weighted mean squared error is used to train the model for accuracy. The proposed approach is checked on the dataset of 2,234 households and the Korea Meteorological Administration data. Two concepts, one using weather data as input and the second using weighted mean squared error function, help predict the power bill. The results indicate that these two concepts reduce about 12.5% of the MAE. The GRU-based model was able to produce results that can be compared with LSTM-based sequence-to-sequence model. The information uses income, size of the house, count of members in the family, and age of members.

Rouibah et al. [5] used IoT in solar photovoltaic (PV) systems for monitoring and evaluating power usage. A low-budget system is proposed for the maximum power point tracking (MPPT), which uses two electronic boards developed by them: one for sensing the acquired data and the second is a converter to enhance the DC-DC. The board which keeps an eye on the power usage includes a board embedded with Arduino Mega 2560 based upon ATmega2560. Also, sensors for sensing current and voltage and a mini BEC voltage regulator and an LCD display were used. An ESP8266 is used as a Wi-Fi module to transmit the required reading and convert in needed data (IPV, VPV, VL, IL, and D) on the Internet. A real-time display on the web is given. The monitored data and real data are compared to make the system improve on accuracy. Any fault detection or sensors failing to transmit data will be intimated as alert messages. Any damage to the system or the remote sensing system is extended research for the future [5].

Roimah Dollah and Hazleen Aris [3] use Big Data Analytics (BDA) to identify consumer's power utilization from sources like smart meters and billing systems [6]. Data on the consumption pattern and trend is identified. The awareness of power consumption helps in saving 20% of energy. The BDA model for tracking power usage is based on BDA layers, which use descriptive analysis and prediction of the data for the consumers [7]. Consumers can view, track, compare, and plan their electricity consumption from home.

9.3   OBJECTIVES OF THE POWER MONITORING SYSTEM

- Identify units consumed by the electronic devices
- An optimization approach to efficiently use power based on consumers behavior
- Near to real usage calculation based on historical data and current data.
- Identifying maximum usage period and devising a strategy to optimize it.

Energy-saving by marking the increase in power consumption and marking the low and high usage.

- Visualization of the usage by plots and charts helps in knowing the usage at one look.
- Any abnormal behavior or malfunction of the system can be easily detected.

Trivially only used power monitoring is done and prediction is at the minimal. Only certain IoT-enabled devices have been programmed to control the flow of electricity and power consumption.

- Monitoring helps keep track of energy consumed daily to alert you to any abnormal behavior.
- Power monitoring is possible on the go with chatbots' help, which can provide power usage whenever asked for or through an SMS.
- Many cities in India use digital energy meter, although many villages still use an electro-mechanical system for reading.
- A home automation system will help in high accuracy bill prediction.
- All electronic devices can monitor current consumption, voltage details using a voltage sensor also possible.
- A genetic algorithm approach helps predict usage with apt crossover and mutation methods, thereby giving better accuracy.
- A low-cost system with highly secured data storage at the cloud to trigger the timestamp interval is also achieved in the proposed research.

9.4   EXISTING SYSTEM

- In the existing Power Monitoring System, only power consumption data are recorded.
- Few IoT-enabled devices can control electricity and power consumption flow, which does not provide the required accuracy.
- Tracking power consumption is not always accurate as no strategy has been proved foolproof.
- The users need to log in to the website only during certain hours, although many apps allow the facility to log in to the same site provided by the electricity board.
- The present system of energy metering and billing in India uses electro-mechanical and somewhere digital energy meters.

9.5   POWER MONITORING

IoT is now widely used by almost all households of all levels and has become a part of the modern world; the significance and utilization are increasing every day. To design efficient and real-time wireless networks to monitor electrical appliances' power consumption, sensors are set at the heap to ascertain current. A circuit is utilized to figure voltage, and with these two, power can be computed. Control qualities are put away in the cloud database. The noninvasive current transformer sensor receives the input from the live power supply and transfers the value to the voltage sensor. Nodemcu gets the value from the voltage sensor and stores it in the cloud database. With the help of a cloud database, users can monitor the power consumption value. Home automation can be achieved with the help of the Relay Module and separate Nodemcu. Devices can be controlled from a cloud dashboard. Power Monitoring System will be very useful for the average salary people and also it overcomes the electricity theft, unwanted power wastage, avoidance of high power consumption devices. Power Monitoring is a term used to monitor the power daily, weekly, and monthly. Electricity is a resource and acts as input for so many gadgets. Most of the world's population is covered by middle-class people. They face electricity charges as a financial barrier, so power monitoring will help us determine the power theft, abnormal bill charge, etc. By monitoring the power, it will also keep the data accurate. Advantages of power monitoring systems are

- By monitoring the power, we can keep the data of power consumption regularly for future reference.
- Future bill predictions can be made.
- Data grasped by power monitoring can be plot on GUI for easy reference.
- Alarm in power monitoring will help people to remind that it is exceeding 500 units.
- An abnormal flow of power can also be detected.

9.6   INTERNET OF THINGS (IoT)

Every activity in our day-to-day life accompanies the Internet, which is nothing but connecting the people for communication, transferring information, and improving the knowledge are some of the available resources. Then the Internet of Things means connecting the physical things with some software and sensors to do some monitoring or gathering information process is possible over here. With the help of the Internet of Things, people can use embedded systems, automation by deploying the IoT with the combination of some ideas on artificial intelligence and machine learning.

IoT is one of the greater revolutions among industries. The main aim of IoT is to combine sensor devices, wireless networks, artificial intelligence, wireless networks, and big data to measure and analyze industrial processes. The IIoT is also the new revenue for businesses which helps in making workflow from starting to end. The major benefits of IoT are discussed as follows, which makes the environment smart. Home environment benefits are now like Amazon product Echo and Google Play will help to play music and proceed with timer and information. Not at all entertaining products also security systems make easier to monitor the temperature, obstacles, etc., public sensors like parking systems, sensors to detect the polluted environment. Many of the cities are changing to smart cities with the help of sensors.

Security, the smallest question on IoT, although comprises encryption and decryption of data while transmitting. Some of the IoT devices have very poor security. There is a risk of hackers found in some IoT devices, so go ahead with some more security options while transmitting the secured data.

9.6.1 Applications of IoT

- Medical and Healthcare
- Transportation
- V2x communications
- Building and home automation
- Industrial applications
- Manufacturing
- Agriculture
- Infrastructure and applications
- Energy management
- Environmental monitoring
- Military applications
- Product digitization

9.6.1.1 Big Data Analytics

BDA is used to analyze a huge amount of data by deploying some algorithms or using some techniques. There are several tools for BDA as follows:

- Xplenty
- Microsoft Power BI
- Microsoft HD insight
- Skytree
- Talend
- Splice Machine
- Spark
- Plotly
- Apache SAMOA
- Lumify

9.6.1.1.1 Xplenty

This is a cloud-based tool to automatically visualize the data flowing automatically through different source places and end places, a platform for transformation, and also a powerful tool. Xplenty's powerful on-platform transformation tools allow you to clean, normalize, and transform data while adhering to compliance best practices.

This pulls the data from different sources, provides much more good security, and provides data to a wide variety of data warehouses, databases, etc. It consists of User Interface by bridging Xplenty API.

9.6.1.1.2 Power BI

This is also a tool for analytics to take data from different sources and change it into an active process. This supports Query Languages like Power Query, Structured Query Language, which consists of user interaction dashboards in build visualizations. But the level of security is low.

9.6.1.1.3 Microsoft HDInsight

This is most probably used in enterprise level organizations to process a huge amount of data and provide Spark service and Hadoop services. There are two things: standard and premium. At the enterprise level, security is provided for developers and scientists, and it provides high productivity by charge fixing and no need to purchase the hardware.

9.6.1.1.4 Skytree

This is one of the most extensive data tools for analytics and data processing and is more accurate and fast. Prediction models are also very easy to use. A variety of algorithms help data scientists with visualization and decision along with interpretability options also. Then the GUI programmed by java is very easy to deploy. Problem prediction data is useful.

9.6.1.1.5 Talend

This is also one of the big softwares for analytics, automating big data. It consists of some graphically designed wizards to accelerate native code generation, allows integration of databases and master database, checking the quality of data, reduces the time consumption on projects with big data, and simplifies using Map Reduce and Spark done by native code generation. Also, Natural Language Processing and machine learning algorithms help to achieve quality data. Agile speeds up data processing.

9.6.1.1.6 Splice Machine

This is also a big analytics tool for big data, comprised of clouds like AWS, Google, etc. A dynamical scale can be deployed to enable the applications; queries can be evaluated automatically distributed in regions and tend to minimize the risk and are deployed very fast.

9.6.1.1.7 Spark

It is the most powerful open source for analytics in big data. It will be very suitable for parallel processes and work with various organizations and large datasets to be processed. It helps an application run in Hadoop faster than 100 times and faster ten times on memory and disk appropriately. The Spark is an open source tool for fast processing and helps in integrating the current Hadoop data with existing Hadoop data. It also consists of some APIs in python.

9.6.1.1.8 Plotly

This is also one of the big data analytical tools which make the data analysis in the format of statistical representation like creating charts for data comparison and share the information on the dashboard of Plotly. By making this representation, it becomes impressive and graphics very informative. It provides delicate information on the basis of data and hosting the file is also provided unlimitedly.

9.6.1.1.9 Apache SAMOA

It is also one of the analytics tools for big data, which helps in the evolution of ML algorithms and also provides various algorithms for the data that has been commonly used.

9.6.1.1.10 Lumify

It is used for visualization and also analysis helps to introduce connections of data and relationships among the data using analytic options. It consists of 2D, 3D graph visibility on layout changes automatically. It also provides the detailed entities on the graph representation. Along with this there are also images and videos. Another feature allows the setting of the project with organized work and workspaces.

9.6.1.1.11 Elastic Search

This is a big data search based on JSON. An analytical engine solves several use cases. Scalability achieved in this tool makes the management very easy, different types of searches like geo search, structured search, etc., and builds many client languages like java, groovy, and python. Real-time search is also possible, along with security.

9.6.1.1.12 R Programming

It is a language used for graphics; various varieties of statistical tests have been used over here. It also consists of the facility for storage and handling the data; a hard copy of it has also been maintained.

9.7 ARTIFICIAL INTELLIGENCE

Artificial intelligence is nothing but human intelligence mimic by machines that performs some tasks. This is mostly used for automation purposes and structured tasks, which also lowers energy costs and makes the machines behave intelligently with some of the computer programs. Some approaches analyze the environments and do some tasks up to some extreme to achieve the energy target. This kind of machine learning helps recognize speech and language in Apple iPhone and supports voice recognition and vision recognition to suggest the products. Now with the advances in technology, the already used benchmarks defined that artificial intelligence is getting outdated, which evolves in reaching major benefits among different industries – widely classified into strong and weak artificial intelligence related to more complex systems. Without personal interaction, the problems can be solved using certain algorithms. Most probably the weak artificial intelligence is deployed in gaming systems and personal assistance like Alexa. Strong artificial intelligence is deployed in self-driving cars and complicated programs.

9.8   CLOUD STORAGE

Cloud storage means storing the data in a remote location and accessing the data without the implementation of software. Nowadays, using the programing language, developing software, and creating website creation, App development is easy because cloud storage does not need to worry about software installation. The software, applications, and websites developed can be accessed from anywhere.

A traditional database needs a separate database for separate concerns and tends to large storage space, but the storage space will be less in the cloud. Anyone having the privilege of accessing data can use the data.

9.8.1   UBIDOTS CLOUD SERVICE

Ubidots offer this type of service; we can convert our sensor data into information through this. They provide APIs to transfer the data securely. This supports three things as follows:

- Ease of use
- Support
- Fairness

9.8.1.1   Ease of Use

The architecture design will focus on the efficiency of data and user experience. With the help of this, users are able to build IoT applications with ease of use which can handle the UX infrastructure.

9.8.1.2   Support

Ubidots mainly focused on providing the good and right tools to make our imagination succeed. They considered our data more valuable.

9.8.1.3   Fairness

Along with both a support system and ease of use, the resource belonging to Ubidots gives information from simple prototypes to production information. At the time of requested development, pricing will become a question, so it consistently achieves the right solution.

9.8.1.4   Sensors

Sensors are nothing, but they are devices used to detect changes in the surroundings like temperature, sound, vibrations, objects, etc. In the day-to-day world, every object has some sensors attached, even the kid playing remote car with sensor devices to find the obstacles on the way car is raveling. Healthcare also has sensors;

organizations have sensor devices like fire detectors, parking sensors, and temperature sensors. In our normal family, the air conditioner has a temperature sensor, for detecting the environment climate and switching on or off the air conditioner automatically. Here, the power monitoring system uses some sensors to detect the power monitoring with the help of monitoring. Sensors usage is also increasing, and it is widely being used in home appliances, shopping mall, hospitals, educators, etc. However, some security-related issues are still prevailing in these sensor devices. This power monitoring system sensor plays a vital role in fetching the live current and conversion; after that converted value will be stored in the cloud database.

9.9 ARCHITECTURE FOR THE PROPOSED POWER MONITORING SYSTEM

The architecture diagram shown above establishes the need for flow control for the transformer sensor to sense the current flow from the live wire and sends an analog input. Current sensors process the analog data to digital form, which forwards to the Nodemcu. The data that reaches Nodemcu is automated to get the data from the current sensor and sends the data to the database. A four-channel relay board that controls all four devices is connected. Nodemcu is connected with the relay module and to our cloud database. Our cloud database consists of a dashboard that displays the power consumption reading and virtual switch to control the devices connected to the relay module. Here the cloud database plays a vital role (Figure 9.1).

The current 30 A is measured by current transformer 30 A, which has been clamped, and the current passing is monitored. Thenodemcu ESP8266 is a Wi-Fi-enabled chip used to transfer the data to a database. Current sensor 30 A senses the current flow. The relay module is used to control power on or power off for the electronic device.

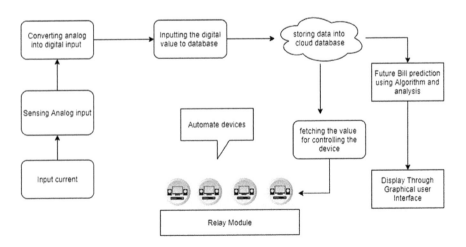

FIGURE 9.1 Architecture diagram for proposed Power Monitoring System using IoT.

9.10   SENSOR COMPONENTS USED FOR POWER MONITORING (FIGURES 9.2–9.5)

FIGURE 9.2   Current transformer sensor 30 A.

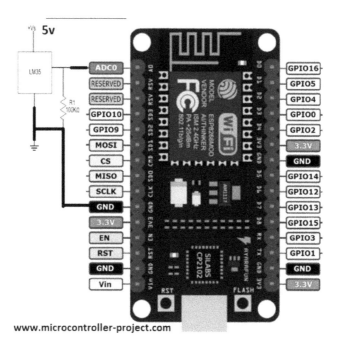

FIGURE 9.3   NODEMCU ESP8266.

FIGURE 9.4   Current Sensor 30 A.

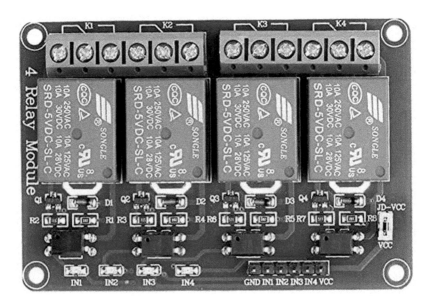

FIGURE 9.5   Relay Module.

FIGURE 9.6 Screenshot of Ubidots cloud database dashboard.

Figure 9.6 represents the power consumption of the device and also the voltage consumption of the device.

9.11 FUTURE BILL PREDICTION

Prediction of the future bill will lead to avoidance of high usage of electricity. The future bill prediction can be made using different comparison ways. One method of energy consumption can be predicted by knowing the maximum energy consumed per day and calculating the billing period. The energy consumption of home appliances can be monitored by measuring the usage of these appliances. For example, if a thousand watts bulb is used for 1 h, it consumes 1000 W for an hour, which is 1 kWh (kilo Watt-hours) equivalent to 1 unit. A normal fan consumes 80 W; an iron box consumes 1400 W; a laptop consumes 100 W; a television consumes 120 W, a water pump/motor consumes 800 W. It can vary from equipment to equipment and brand to brand. Another method is to compare with previous year's same period usage and predict the bill based on that. This gives more or less appropriate consumption if new appliances have not been added. Now you will get one valid answer, yes now the current month's bill has been predicted. The second means looks back to the last year's June month data 2019 by comparison; also the data can be predicted.

Figure 9.7 illustrates the entire power monitoring system phases: the first phase demonstrates the power monitoring system. The second phase comprises future bill prediction and is further classified into calculating upcoming electricity bills and

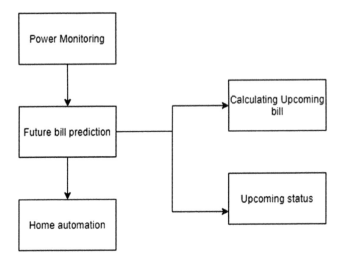

FIGURE 9.7   Automated Future Bill Prediction diagram.

summary status. The third phase comprises home automation. Power monitoring is done by sensors and stores the data into a cloud database. With the help of data in the database, future bill predictions like summary status and the upcoming bill can be calculated. We can achieve home automation by combining device information with the relay module.

According to future bill, consider the real-time application of electricity bill calculation for one year, a total of 12 months. The electricity bill will be charged once after two months of completion. So for January and February, the bill amount may be b1; next for March and April month, the bill amount may be b2, which will range from b1, b2, b3, b4, b5, and b6 for 12 months, respectively, for year y1. Now for next year y2, the bill amount is b1, b2 ....b6. Surely there will be a relationship between y1(b1) and y2(b1), so the predicting amount for year three i.e., y3(b1)>y1(b1)|ly2(b2) now the resultant bill amount for year3 first bill amount become either b1 of y2 month or b1 of y1. Hence, the result becomes R1.

The next simultaneous method is to consider that the electricity bill will be charged once in 60 days now while you are in an intermediate day, i.e., you can randomly pick any one of the days. Here, I consider 45th day of billing date, means now consumption of power is nearly 345 units, so calculate no of unit consumption per day that becomes d= 1day= no of unit consumed per day, afterward for 45 days =d*45 as like that 60 days 60 *d. The result becomes S1.

Combining the two resultant values, i.e., R1 and S1, the accurate value can be found and becomes an accurate, optimized value. Just pick the accurate value, i.e., which one charged the accurate value, and opt for the future bill prediction system (Figure 9.8).

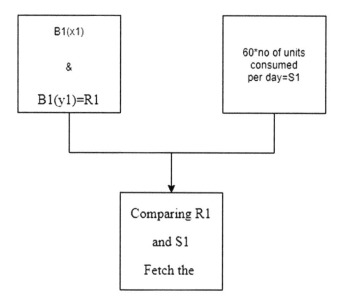

FIGURE 9.8   Automated Future Bill Prediction diagram.

9.12   CONCLUSION

By using the Power Monitoring system using IoT and future bill prediction, efficient usage of power consumption will be taken. It will reduce the financial barrier for people getting moderate wages and less wages. So many efficient ways are there to predict the future bill, but need to pick the efficient one matters here.

REFERENCES

[1] T. Zhang, H. Shi, M. Chen, L. Xue, Y. Hao and J. Qu, "*Design of Intelligent Power Load Automatic Monitoring System,*" in: *The 2019 IEEE 3rd International Conference on Electronic Information Technology and Computer Engineering (EITCE)*, Jimei University, Shanghai University of Engineering and Technology, Guangdong Academy of International Academic Exchange, Xiamen, China, October 18-20, 2019, pp. 850–854, doi: 10.1109/EITCE47263.2019.9095036.

[2] H. Tavares, B. Prado, K. Bispo and D. Dantas, "*A Non-Intrusive Approach for Smart Power Meter,*" in: *2018 IEEE 16th International Conference on Industrial Informatics (INDIN)*, 2018, pp. 605–610.

[3] R. Dollah and H. Aris, "*A Big Data Analytics Model for Household Electricity Consumption Tracking and Monitoring,*" in: *2018 IEEE Conference on Big Data and Analytics (ICBDA)*, 2018, pp. 44–49, doi: 10.1109/ICBDAA.2018.8629769.

[4] J. Park, M. Kim, S. Hong, Y. Jeong and I.-H. Suh, "*Prediction of Individual Household Energy Bills Using Deep Learning,*" in: *The 35th International Technical Conference on Circuits/Systems, Computers and Communications*, July 3–6, 2020, WINC Aichi, Aichi Industry & Labor Center, Nagoya, Japan, pp. 500–504.

[5] N. Rouibah, L. Barazane, A. Mellit, B. Hajji and A. Rabhi, "*A Low-Cost Monitoring System for Maximum Power Point of a Photovoltaic System Using IoT Technique*," in: *2019 International Conference on Wireless Technologies, Embedded and Intelligent Systems (WITS)*, 2019, pp. 1–5, doi: 10.1109/WITS.2019.8723724.

[6] H.K. Temraz, M.M.A. Salama and V.H. Quintana, "Application of the Decomposition Technique for Forecasting the Load of a Large Electric Power Network," *IEE Procs. Gener. Transm. Distrib.*, vol. 143, no. 1, 1996, pp. 13–18.

[7] S. Hosein and P Hosein, "*Improving Power Generation Efficiency Using Deep Neural Networks*," in: *Proceeding of the 2016 ICML Workshop on #Data4Good: Machine Learning in Social Good Applications*, pp. 1–5.

10 Cognitive Computing-Powered, NLP-Based, Autonomous Question-Answering System

Atul Prakash Prajapati and D. K. Chaturvedi
Dayalbagh Educational Institute, Agra, India

CONTENTS

DOI: 10.1201/9781003156123-10

10.1 INTRODUCTION

Question-answering (Q-A) systems have been a scorching area of study. The academic and industry personnel have proposed a lot of approaches over time. Thus, plenty of systems have been designed and introduced over time, like ITS (Intelligent Tutoring Systems), Expert Systems, and Simulation tools (for offering real-world hallucination to novice users). The fundamental idea of the Q-A system is to offer concise and precise answers to the user query rather than the list of documents. Early search engines were providing keyword-based searching which bestows the list of documents and users have to squander a large amount of their time and energy in finding out the exact answers to their queries. Thus with the advent of technology, various techniques have been anticipated by the researchers for the improvement and utilization of the digital data. Therefore, the following section gives a concise introduction to the systems proposed under Q-A domain.

Objectives of the chapter are as follows:

- Implementation of the "QAEdu" architecture, a noble cognitive agent's architecture in the field of education.
- Implementation of cognitive computing techniques (NLP, Knowledge Base, Expert agent System, etc.) within a single proposed architecture for enhancing the efficiency of a Q-A system.
- Implementation of recommendation system and Confidence Value concepts for the authenticity of provided answers by the agent.
- Implementation of Experiential Knowledge-Base.
- Implementation of the cognitive agent-based Q-A system for the edification of graduate students by teaching basic "electromagnetic" concepts.

With the advent of technology, various techniques have been anticipated by the researchers for the improvement and utilization of digital data. In the early era, 1990 Tim Berner lee had projected the idea of "Web". This early Web (Web 1.0) has lots of restrictions like if you desire to publish your contents on the Web, you have to seek permission from the network operator (this is for security aspects). Further web 2.0 endowed users with the freedom of expressiveness. At that moment the volume of data was continually increasing, and also people were dealing with relational databases and schemas (traditional databases were structured in nature) [1–6]. At this point, one has raw data that needs to be analyzed for building assumptions and also there is a lack of semantics in the data (because of its structured nature). These systems have well-defined language syntax and structuring (SQL). Subsequently, with the advent of web 3.0, things have changed significantly at this time one has lots of data, and also there was no restriction on the structure of data. People are dealing with all types of data (Structured, Semi-structured, and Unstructured). It has endowed with the ability to access web information very easily. Still, there was a problem of finding exact/precise/accurate answers from such a large pool of data and people were using the search engines for extracting the knowledge from the web. These search engines were providing the list of documents rather than the direct answers. It leads the researchers in the direction of Q-A systems, which should have the capability of providing immediate answers to the user's questions. Therefore, the academic and industry personals had proposed quite a lot of approaches over time. Plenty of systems have been designed and introduced over time, like ITS, Expert Systems, and simulation tools for offering real-world hallucination to novice users. The fundamental idea of the Q-A system is to provide concise and precise answers to the user's query rather than the list of documents. Early search engines were providing keyword-based searching which bestows the list of documents and users have to squander a large amount of their time and energy in finding out the exact answers to their queries. This was a time-consuming task; therefore, researchers proposed lots of systems from time to time to overcome such problems. Based on the domain specification, Q-A systems have been categorized into two types: Closed Domain and Open Domain systems. Researchers stimulated toward either Open Domain systems (web mining approach-based systems) like [7–9] or Closed or restricted Domain systems [10–12]. These two different domain Q-A systems can be categorized based on knowledge representation and extraction because it is assumed that a good understanding of domain is much better than a suitable algorithm. So the center of attention is the construction of practical knowledge base. For that, various techniques have been proposed over time (like Frames, Scripts, Databases, etc.). Further, a new era of systems has been started contrary to NLIDB systems; now people are more interested in the semantically searching of data, so various techniques like Semantic Network [13] and Ontology [15] concept has been proposed for the construction of knowledge bases. The fundamental idea behind the semantic network was to offer some inherent sense to our data. So, a machine can semantically process the data and can help us in better inferencing. However, it failed to serve its essential purpose. Although its name is a semantic network, it does not covenant with semantics anyhow. It was following the XML/RDF structure (XML follows "Tree" data model, and RDF supports XML, N3 (Notation-3, which has (Turtle and N-triples as its subset)), and JSON-LD, etc.). XML itself is not self-explanatory, one need DTD (document-type definition) or schema to recognize the significance of XML. Means it is dealing with the syntax rather than semantics. It spotlights much on the structure of data (how one should structure or organize our data). Here the focus is on "How", rather than "What" (What is the meaning of our data). Subsequently, the ontology has changed conventional thinking. Though it is using the markup languages for its knowledge modeling, it has provided some additional features to its structure. Contrasting RDFS, in ontology, people are using the concept of classes, subclasses, properties (Reflexive, Symmetric, transitive, etc.). It has provided additional meaning to our data, and it is domain-specific also. It models the meticulous domain of interest (taxonomic approach). It uses the notion of facts (triplet combination Subject, Predicate, and Object) for knowledge modeling [15–30]. However, ontology represents a closed domain knowledge base where individual models the domain information using web semantic languages. It symbolizes the domain or modules the domain knowledge provided by the group of experts. In addition to ontology, there is another side where the systems dig out knowledge from the Web (unstructured data), these systems follow the concept of Information Retrieval (IR) systems. Such systems use the concept of "Term Frequency/Inverse Document Frequency" (Tf/Idf) (term weighting for the selection of best sentence/page). The first web-based Q-A system was "START" [31], started in 2004 at MIT. Subsequently, numerous systems have been proposed [32–36]. Q-A systems are also classified based on various approaches used for question understanding and answer extraction like Linguistic approach, Statistical approach. To comprehend user natural language question and to retrieve the accurate and precise answer, the linguistic approach uses various NLP functions (Tokenization, POS Tagging, etc.). These systems use NL frontend and use conventional database queries to retrieve the answers from the structured database. If one commences from the old systems BASEBALL: An automatic Q-A system [1, 37] was the first system based on this approach. Subsequently, several other systems have been proposed [32–36]. Contrary to the linguistic approach, the statistical approach is different from the relational database concepts. It deals with the large and heterogenic unstructured data. Such systems require a massive amount of data to learn (training) further it make the model out of this and apply the produced hypothesis to the new domains. These techniques are used at different stages in Q-A systems. Some of the approaches akin to Maximum Entropy Model, Support Vector Machine, and Bayesian Classifiers have been used for the classification of questions [38]. The maximum entropy model uses for classification. It uses a bag of words or N-Gram method for this purpose [39–41]. Support Vector Machine method is also used for the question classification. A new era started in the area of Q-A systems, Visual Question Answering (VQA) systems. Such systems combine the concept of Computer Vision and NLP (Natural Language Processing), which uses image-based information [42–47] and proposes "iTutor" for graduate students for instructing basic skills of word processing, spreadsheet, presentation graphics, and webpage production and some theoretical models for an acquaintance in the course of ICT, such as the domain model (as the problem-solving situation model), the student model (as the interaction and process model), and the teaching model (as the feedback model). They used learning materials like prior knowledge assessment and comprehensive assessment, Research design and procedures (a quasi-experimental design), and data collection and analysis. Hooshyar et al. [48] proposed FITS (a flowchart-based intelligent tutoring system for improving problem solving skills of novice programmers) for novice programmers to tutor essential flowchart-development concepts (Computer Programming). They used Bayesian networks for the process of decision-making, multi-agent system and an automatic text-to-flowchart conversion approach. Harley et al. [49] Authors proposed an ITS system for undergraduate students to teach complex science topic. They used the MetaTutor, a multi-agent digital learning environment. In this system, automatic facial expression recognition, self-report, electrodermal activity techniques are used as a part of AI techniques. Grawemeyer et al. [50] proposed a iT alk2Learn system for the children aged 8–12 years to tutor them basic fraction concepts (mathematics). In this system, Bayesian networks, performance only (non-affect condition), and intelligent affect-aware support techniques are included in a learning platform as a part of AI methods. ElGhouch et al. [51] proposed an ALS-CORR [LP] system for the bachelor of computer science students to educate them the Merise Method concepts [used in Conceptual Data Model (CDM) for crafting the databases]. By the system ALS-CORR [LP] Adaptive learning system with a Correction of Learning Path is used as an AI technique. Grivokostopoulou [52] proposed an Educational System for schooling students to teach them searching algorithms, an artificial intelligence topic. Topics like Algorithm Visualization, Interactive Exercises, Example Cases (for practical implementation), Assessment Approach (for performance evaluation), feedback to students, similarity between the students and the correct answer using the edit distance metric, type of the answer, and examples of errors are identified by the system and the response is automatically marked via an automated marker. Samarakou [53] proposed an ITS system Stu-DiAsE (Fuzzy-logic-based model used as an AI technique) for engineering students to teach Foundation of Energy, Operation of Local Computer Networks, C Programming Language, Energy Management, and introduction to MATLAB concepts. Mostafavi et al. [54] proposed an ITS system DT4 (DeepT hought-4) system for computer science students to teach them deductive logic concepts which are explained in computer science education. Data-driven tutoring, on-demand hints and worked examples, DDML system, DDPP approach is used as an AI technique. Masoud et al. proposed a SQLTutor for students who want to learn database concepts as a part of Software Engineering. Intelligent Tutoring System Builder (ITSB), Student Module (SM), Pedagogical Module, Domain Knowledge, Communications Module, and assessment of the student performance concepts are used as a part of system implementation techniques [55]. Graesser et al. proposed an ITS system Electronix-Tutor for the students who want to learn STEM topics. Electronix Tutor teaches mathematics, electronics, and dynamical systems. SuperGLU (Generalized Learning Utilities) open-source framework used as one approach for handling the problems of integrating and coordinating distributed web-based systems in real time. My Scores tab is used as student feedback tab. The concept of Recommender System is also used in Electronix-Tutor system. It integrates multiple pre-existing ITSs and traditional learning resources (like Auto Tutor, Dragoon, Learn Form, BEETLE, Assessment's, Point and Query, readings, and videos) into a coherent user experience [56]. Prajapati et al. proposed a cognitive system SN: CQA to help Engineering Students (Electrical Engineering) in learning and understanding the Basic Electrical Motor Concepts. NLP, Semantic Network, and cognitive computing techniques are used as implementation of cognitive computing techniques for practical implementation of a robust Q-A system [57]. In most of the cases, the erudition is performed in abstraction, thus these systems are designed for the specific and restricted jobs, but these days with the dawn of technology there is a need of such Q-A system that can at least mimic humans. Thus, people are talking about cognitive computing-based system construction approach which proposes an architectural way of designing the systems. A cognitive Q-A system is defined as "Systems that can sense, comprehend and act". It assumes that these systems should mimic the humans while giving the exact, precise, accurate, and the best answers to the user queries. Most of the early systems are either ornamenting a particular technology or suggesting new approaches in this area. But very few are talking about the architectural way to system crafting. If we want to put into practice a dynamic cognitive Q-A system, we have to build an architecture first that can synergize all the techniques of cognitive computing under one roof. The architecture of QAEdu can deal with all the three necessary steps of cognitive computing techniques (Sense, Comprehend, and Act) [58].

- Sensing deals with Computer Vision, Audio Processing, and Sensor Processing. Comprehension deals with implementation of NLP techniques and knowledge-based construction.
- Action implements the concept of inference engine, Machine Learning concept, and the Expert System approach.
- In the proposed architecture QAEdu, we have tried to put into practice all the possible aspects of cognitive computing techniques and will be implementing other techniques shortly.

10.2 BACKGROUND

Earlier period Q-A or ITS, which were premeditated for the deployment of digital information present, were facing abundant problems. A few of them are scalability, knowledge updating, accumulation of latest information, and many others. Subsequently, to surmount these problems QAEdu proposes a novel approach to Q-A systems. It is introducing the cognitive computing, NLP, Knowledge Base, Confidence Value, and Answer Recommendation System-powered Q-A system. At this point, the cognitive computing approach provides features like speech processing, effective knowledge modeling, computer vision, and much more. NLP endows the power of understanding user questions profoundly and accurately, and for a robust knowledge base, we are following (XML) format for accumulating facts in knowledge base) as it is the best appropriate approach for such Q-A systems among all available techniques. We have checked Ontology and Semantic Network methods [13, 15] for knowledge representation but did not find these suitable for Q-A systems that are giving long answers to the user questions. The sections are defined as follows:

1. Structure of the present research work is as follows:
   a. Implementation of Concept Model.
      i. Architecture of QAEdu.
      ii. Question understanding.
         - Analysis of question type.
         - Semantic Component (Entity and Association) extraction from the question.
         - Evaluation table for question understanding.
      iii. Designing of Natural Language Processing Corpus.
         - Designing of NLP corpus modules.
         - Role of attributive searching (Adjectives etc.).
         - Role of dictionary concept for finding and replacing synonyms.

b. Implementation of Experience and Knowledge Model.
    i. Knowledge-based construction.
       • Extractions of facts from documents.
       • Evaluation of the extraction process of the facts from the provided documents.
   ii. Answer searching and providing the best and recommended answers.
       • Based on Confidence Value (For finding the best answer).
       • Based on user feedback (Liking Values for answers) (For Suggested answers).
c. Implementation of Interface Model
    i. Methodology and Experimental Results (Case Study).
       • Flowchart for the explaining the working and construction process of QAEdu.
       • Experimental Results.
d. Evaluation of QAEdu.
    i. Collected sample questions inspired by TREC (2009–2011) questions set, and evaluated the performance of QAEdu.

10.3 CONCEPT MODELING OF "CONCEPT MODEL", "EXPERIENCE/KNOWLEDGE-BASE MODEL", AND THE "INTERFACE MODEL" OF THE PROPOSED COGNITIVE AGENT "QAEDU"

10.3.1 PROPOSED ARCHITECTURE

For illuminating each module and construction process of this cognitive Q-A system, the following assumptions have been made. This system has TM (Teacher Module), SM, QM (Querying Module), KCM (Knowledge Corpus Module), NCM (NLP Question Corpus Module), RM (Reasoner Module), and DM (Decision-Making Module).

Definition 10.1

QAEdu Q-A system has following tuples (TM, SM, QM, KCM, NCM, RM, DM). Refer to Figure 10.1.

(a) TM – Teacher Module: This module subsists at UIM interface. In this module, a teacher or domain expert provides the necessary documentation for answer extraction and manages the question corpus defined and managed using the (NCM) module. Here, the role-based access function allows the teacher to have both the options "R/W" (Read, Write).

(b) SM – Student Module: This module subsists at UIM interface. In this module, a student accesses the system for asking the query. There exists a "Role-Based" access function, which defines the role to access the system. A student's role-based function allows to have "Read Only" functionality to the student.

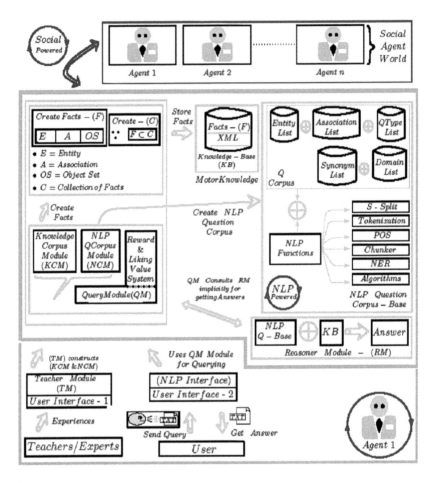

FIGURE 10.1 Architecture of ("CogQA_Edu") a "NLP-based Cognitive Question - Answering System". This system has "TM" (Teacher Module), "KCM" (Knowledge Corpus Module), "NCM" (NLP Question Corpus Module), "QM" (Query Module), and "RM" (Reasoner Module). In "TM" module, a teacher or expert provides the necessary documentation for answer extraction or stuff the facts in knowledge corpus directly and manages the "NLP" question corpus also. "KCM" module is controlled by the "TM" module. Using this module a teacher or an expert fills the facts in knowledge corpus. "NCM" module is also controlled by the "TM" module. Using this module a teacher or an expert fills the entities, associations, dictionary values, and domain identification values in the question corpus. "QM" handles the user queries. It implicitly uses the reasoner module for inferencing and ascertains relationships between entities. This module is implicitly controlled and used by the "QM" module. The above diagram shows the detailed functions of "Agent1" and also represents the society of agents. These agents will be the experts of their respective domains.

(c) QM – Query Module: This module also subsists at UIM interface. It handles user queries. It implicitly uses the reasoner module, decision-making module for inferencing and ascertains relationships between entities. It extracts the best answer from the knowledge base.

(d) KCM – Knowledge Corpus Module: This module has been controlled by the TM module. Using this module a teacher or an expert fills the facts in knowledge corpus.

(e) NCM – NLP Question-Corpus Module: This module has also controlled by the TM module. Using this module a teacher or an expert fills the entities, associations, dictionary values, and domain identification values in the NLP question corpus. These values help the "NLP" functions in the extraction of linguistic components from the user question/documents.

(f) RM (Reasoner Module): This module is implicitly controlled and used by the (QM) module.

(g) DM (Decision-Making Module): It is a part of (QM) module. It helps the (QM) module in finding the answer form the agent's knowledge base. It internally consists of (RM, InF, NLP) modules.

Definition 10.2

Designing of "Role-Based function", "NCM", and "KCM" modules.

(a) Designing of "Role-Based Access Function" for defining the "Roles" of (TM), (SM) modules.

$$RB_f := \exists,\ Usr : \left(Usr \in (T,S)\right)\ \&\&\ \left(Role \in (R,W,RW)\right)$$
$$: \left(T \cdot RB_f \leftarrow (R,W,RW)\ \&\&\ \left(S \cdot RB_f \leftarrow (R)\right)\right) \tag{10.1}$$

There exists a role-based access function which assigns different roles to the users of the system. It has following roles "read, write, read-write", ("r", "w", "rw"). Such that a teacher can have all the above-defined roles, whereas a student has read only ("r") access to the system.

(b) Designing of the TM for the construction of (KCM), (NCM) modules.

$$TM := \exists,\ \{NCM,KB\} : \left\{(KB) \leftarrow (T_i) \cdot (T \cdot RB_f)\right\},$$
$$: \left[T_i \in S\{T \| DEx\}\right]\ \&\&$$
$$\exists, Td_i \leftarrow T_i : \left(Td_i \in \sum_{i=1}^{n} Td_i \equiv \left(S\{Doc\} \leftarrow S\{T\}\right)\right) \tag{10.2}$$
$$\therefore \left[\left(LC \equiv (x_i,y_i.)\right) - \left|(Td_i \cdot NLP_f)\right|\right] \rightarrow \left[[Temp] \in \{E,A,OS\} \equiv F_i\right]$$
$$\&\&\ \left[\sum_{i=1}^{n} F_i \equiv (KB)\right]$$
$$\therefore (T_i \uplus TM) \rightarrow \left[(KB)\ \&\&\ (NCM)\right]$$

There is a TM, which is responsible to manage both the (NCM, KCM) modules, provided that a teacher is the domain expert who provides the necessary documents for knowledge extraction. The knowledge is extracted from the documents using the "NLP" functions and maintains in the knowledge base using a fixed format. This format is called a triple representation or the fact. Therefore, the teacher uses the (TM) module for defining the knowledge base and the "NLP" corpus.

(c) Designing of the SM module.

$$SM := \exists, \{Q_i\} \leftarrow \left((S_i) \cdot (S.RB_f) \right) : \left[S_i \in S\{Student\} \right]$$
$$\&\& \quad \exists, QM_f \ni DM_f \tag{10.3}$$
$$\exists, (ans_i \leftarrow Q_i) \leftarrow \left[(KB) \otimes (QM_f) \right]$$

The (SM) module is used to access the system by the students for asking their questions. These questions are handled by the (QM) module, which internally takes the help of (DM) module for extracting the answer from the knowledge base.

Definition 10.3

The "TM" module has two tuples <KCM, NCM>.

$$\begin{cases} KCM \in TM, & Knowledge\ Corpus\ Module. \\ NCM \in TM, & NLPQ - Corpus\ Module. \end{cases} \tag{10.4}$$

This TM module is accountable for the supervision of both the modules (KCM and NCM). A teacher provides the documents for filling up the facts in knowledge corpus and also sets up natural language processing-based question corpus.

(a) The "Natural Language Corpus Module" (NCM) is defined as follows.

$$NCM :=$$
$$\exists, (NCM,NLP_f) : \left[NCM \ni \{S\{E\},S\{A\},S\{Q\_type\},S\{S\},S\{DIE\},.\} \right]$$
$$\&\& \quad (TM \rightarrow NCM) \quad \&\& \quad (NCM \rightarrow NLP_f) \tag{10.5}$$
$$\therefore LC \leftarrow \{S(Doc) \| S(Q)\} \leftarrow (NLP_f \uplus NCM) \quad \&\&$$
$$LC \rightarrow \left[(Q\_Under - |S(Q)) \right]$$

The (NCM) module helps the "NLP" functions in their working. It maintains the list of "Entities" (E), "Associations" (A), "Question-type" (Q-type), "Synonym" concept for the implementation of dictionary concept (S(S)), and the "Domain Identification Entities" (DIE) list. These lists are maintained and defined by the teacher who helps in extracting the linguistic components (LC) from the user questions. These components increase the understanding power of the agent.

(b) The Knowledge Corpus Module (KCM) module is defined as follows:

$$KCM := \begin{bmatrix} \forall,(E,A)\xrightarrow{\ M\ }_1(O):[O \in OS] & \therefore \\ [(E,A,OS) \to F_i] & \&\& & \left(\sum_{i=1}^{n} F_i \equiv agent(KB_k) \right) \end{bmatrix} \quad (10.6)$$

The knowledge corpus module (KCM) is responsible for maintaining the knowledge base of the proposed agent. Here, we have collected documents from the domain experts regarding the "Electrical Motor" concepts. The domain experts have provided the documents based on different category of "Electrical Motor". Therefore, for each question ((E, A) pair) there exist multiple answers which are represented by the set of "Objects".

Definition 10.4

The KCM module has following tuples <F, C, E, A, OS, Ax>.

(a) Define Fact.

$$F := \{t\} : t = \langle E,A,OS \rangle; \quad (E \in \{Entity\ Set\}), \quad (A \in \{Association\ Set\}), \\ (OS \in \{Object\ Set\}) \quad (10.7)$$

An expert supplies the fact into the knowledge base to symbolize knowledge. The fact is a triplet (entity, association, object set) (for each (E, A) pair we are collecting group of answers provided by the group of teachers).

(b) Define Class (collection of facts, or knowledge base).

$$C := \exists, \quad F_j : \left\{ F_j \in S(F) \equiv \sum_{j=1}^{n} F_j \right\} \to (KB) \quad (10.8)$$

C is a collection of similar facts (Class) that belongs to same entity set and associations but has different objects. The knowledge base is the collection of facts.

(c) Defining the Entity (E).

$$E := \exists, \quad (E \in NP) \in (LC) \leftarrow \left[[(Sen) \in \{Doc\}] \otimes (NLP_i) \right] \&\& \\ (E \in [E,A,OS]) : (E-|(Ele\_Mtr\_Com)); \\ (NP \ni \{N \uplus (Art \uplus JJ \uplus Part \uplus PP)\}); \\ (N \ni (NN,NNS,NNP,NNPS)); \quad (Art \ni (A,An,The)); \\ (JJ \ni (JJ,JJR,JJS)); \quad (Part \ni (Pre\ P,Past\ P,Fut\ P)); \\ (PP \ni (PP\_Abs,PP\_Adj)) \quad (10.9)$$

An "Entity" (E) is an element of "Noun Phrase" (NP). We extract the linguistic components from the sentences of the documents. These linguistic components consist of various elements (NP, VP, ...). The entity is an element of the triple <E, A, OS>. In the proposed system, the entity represents the "Components of the Electrical Motor". Further, the "Noun Phrase, Noun, Adjective, Participle" (NP, N, JJ, PP) concept is defined in Equation (10.9).

(d) Defining the Association (A).
Similarly the association is an element of the "Verb Phrase". It is an element of linguistic components which are extracted from the documents using the "NLP" functions. An association is an element of the fact triple (E, A, OS) and relates two entities.

$$A := \exists, \ (A \in VP) \in (LC) \leftarrow \left[\left[(Sen) \in \{Doc\} \right] \otimes (NLP_f) \right] \ \&\&$$
$$: \left(A \oplus (E^R O) \right)$$
$$(V \ni \langle V.* \rangle \rightarrow VP \ni V(NP|PP)*);$$
$$\left(NP \ni \{N \uplus (Art \uplus JJ \uplus Part \uplus PP)\} \right);$$
$$\left(N \ni (NN,NNS,NNP,NNPS) \right); \left(Art \ni (A,An,The) \right);$$
$$\left(JJ \ni (JJ,JJR,JJS) \right); \left(Part \ni (PreP,PastP,FutP) \right);$$
$$\left(PP \ni (PP\_Abs,PP\_Adj) \right);$$

$$(10.10)$$

(e) Defining the object set (OS).

$$OS := \exists, \ (Q_i,O_i) : (O_i \in NP) \in (LC) \leftarrow \left[\left[(Sen) \in \{Doc\} \right] \otimes (NLP_f) \right]$$
$$\&\& \ (O_i \in OS_i \in [E,A,OS]) : \left(O - |(Ele\_Mtr\_Com) \right)$$
$$\left(\sum_{i=1}^{n} O_i \rightarrow S\{OS_i\} - |Q_i \right);$$
$$\left(NP \ni \{N \uplus (Art \uplus JJ \uplus Part \uplus PP)\} \right);$$
$$\left(N \ni (NN,NNS,NNP,NNPS) \right); \left(Art \ni (A,An,The) \right);$$
$$\left(JJ \ni (JJ,JJR,JJS) \right); \left(Part \ni (PreP,PastP,FutP) \right);$$
$$\left(PP \ni (PP\_Abs,PP\_Adj) \right);$$

$$(10.11)$$

An object is the element of "Electrical Motor Components". It is the "Noun Phrases" which are extracted from the sentences of the documents. For each question there exist many objects/answers which collectively belong to the object set for that particular question. This is because of the fact that we have lots of documents available for the same topic provided by the teachers.

(f) Define Axioms (Rules).

$$A_x := \left\{\left(A_i, rel\left(E_j, O_k\right)\right)\right\} : \left(i, J, K\right) > 0; \tag{10.12}$$

A_x is a set of axioms. These are the rules that tie the components of a triple semantically.

Definition 10.5

The "NCM" module consists of the following tuples: $\begin{pmatrix} N_{SE}, N_{SA}, N_{SDE}, N_{SQT}, \\ N_{SS}, N_{SFN}, N_{SAlg} \end{pmatrix}$

(a) Define "Set of Entities".

$$N_{SE} := \exists,\ NE_k : \left[NE_k \in (E, A, OS) \equiv F \right] \quad \&\&\quad NE_k \in \sum_{i=1}^{n} NE_k \tag{10.13}$$

N_{SE} is a set of entities that are stored in the "NLP-Q-Base" entity list.
(b) Define "Set of Associations".

$$N_{SA} := \exists,\ NA_k : \left[NA_k \in (E, A, OS) \equiv F \right] \quad \&\&\quad NA_k \in \sum_{i=1}^{n} NA_k \tag{10.14}$$

$$\left(NE_i \uplus NA_k \uplus NE_j \right); \left(NE_i, NE_j \in S\{NE\} \right)$$

(N_{SA}) is a set of associations that are stored in the "NLP Q-Base" association list.
(c) Define "Set of Domain Identification Elements".

$$N_{SDE} := \exists,\ (Q, NE) : \left[NE \in S\{DIE\} \right] \quad \&\&\quad \left[Q \in S\{Usr(Q)\} \right]$$
$$iff,\ Q \ni \left(S(NE) \in S\{DIE\} \right),\ \ \therefore ans_k \leftarrow \left[(SM \uplus QM_f) \otimes (KB) \right]$$
$$Otherwise,\ \left[SM \rightarrow Out\ of\ Domain\ Question \right] \tag{10.15}$$

(N_{SDE}) is a set of domain identification elements that are stored in the NLP Q-Base. For each question, we are going to extract the linguistic compo-nents and finally check by using these linguistic components that whether the question belongs to the "Electrical Motor" domain or not. If the ques-tion belongs to the expected domain, the agent will fetch the best answer otherwise it will return an error message.
(d) Define "Set of Q-type Elements".

$$N_{SQt} := \exists,\ Q-type : \left[Q-type \in WP \right] \quad \&\&$$
$$Q-type \in \sum_{i=1}^{n} Q-type \equiv S(N_{SQt}) \tag{10.16}$$

(N_{SQt}) is found out by using the combination of set of interrogative pronouns (WP) and associations. Subsequently, this combination is used to classify question type.

(e) Define "Set of Synonym Elements" for implementing the dictionary concept. (N_{SS}) is a set of synonyms that advances the question understanding method. It also provides the attributive searching during the linguistic component extraction process.

$$N_{SS} := \exists, \; S_k : \left[\left(Sem(S_k) \equiv Sem(A_m) \right) \| \left(A_m \ni S\{S\} \right) \right]$$

$$\&\& \; \left(S_k \in \sum_{i=1}^{n} S_i \right)$$

(10.17)

(f) Define "NLP Functions" for extracting the Linguistic Components.

$$N_{SFN} := \exists, \; Td_i : \left[Td_i \in \left(Doc \| Que \right) \right] \; \&\&$$

$$\exists, \; NL\_Fun : \left[NL\_Fun \ni \left(S.Split, S.Tok, S.POS, S.Chunk, etc. \right) \right]$$

$$\left(X \in \left(x_i, y_{i\cdot} \right) - \left| \left(Td_i \uplus NL\_Fun \right) \right) \; \&\&$$

$$X \rightarrow Analytical\_Processing$$

(10.18)

(N_{SFN}) is a set of NLP utilities, which consists of following features (Split Sentence, Tokenization, POS (parts of speech tagging), Chunker, EAR (Entity, Association recognition), etc.). These functions help in extracting the linguistic components from the documents/questions which are subsequently used for various analytical processing purpose.

(g) Define "Algorithms" for implementing above-defined functions.

$$N_{SAlg} := \left\{ \sum_{p=1}^{n} NAlg_p, \forall \left(n > 0 \right) \right\}$$

(10.19)

This is a set of algorithms that are used in the implementation of NLP functions and in mining the essential values.

Definition 10.6

Query Module (QM) consists of the following four tuples:
$\langle Q, CLVM, SM, DM, \; (NCM, KCM) \in RM \rangle$

$$\begin{cases} Q = Input \; Question. \\ RM = It \; implecitely \; uses \; the \; both \; (NCM \; and \; KCM). \\ TM = If \; the \; value \; is \; not \; found \; in \; KC, \; TM \; module \; is \; in \; formed \\ for \; the \; required \; action. \end{cases}$$

(10.20)

(a) Confidence and Liking Value Module (CLVM) consists of the following tuples:

$$\langle f(L), f(Cof), f(AnsL), f(AnsCof), f(AS), f(AB) \rangle \qquad (10.21)$$

(b) Define "Like Value" and "Liking Value Function" for implementing the suggested answer concept.

$$f(L) := \left\{ \sum_{i=-10}^{10} l_i, \forall i = (-10, ., 0, ., 10) \right\} \qquad (10.22)$$

$$
f(AnsL) := \Big(\exists, \ ans_i : \big[ans_i \leftarrow S\{ANS\} \in (KB) \big] \Big)
$$
$$
\&\& \ \Big(\exists, Usr_j : \big[Usr_j \leftarrow \big(S\{USR\} \equiv S \big) \big] \Big) \qquad (10.23)
$$
$$
: Usr_j \rightarrow \left(\forall \sum_{i=1}^{n} ans_i \in Q_i \right) \otimes f(L)
$$

(L) is a set of Liking Values. Here user assigns a particular value from the range (−10,.,0,.,10) to each answer, based on his/her liking for the answer. (AnsL) is a function, which links Liking Value to each answer of the Fact. This is used in the process of answer suggestion.

(c) Define "Confidence Value" and "Confidence Value Function" for defining the best answer.

$$f(Cof) = \left\{ \sum_{i=0}^{10} Cof_i, \forall i = (0, ., 10) \right\} \qquad (10.24)$$

$$
f(AnsCof) := \Big(\exists, \ ans_i : \big[ans_i \leftarrow S\{ANS\} \in (KB) \big] \Big)
$$
$$
\&\& \ \Big(\exists, \ ExR_j : \big[ExR_j \leftarrow \big(S\{ExR\} \equiv DExR \equiv T \big) \big] \Big) \qquad (10.25)
$$
$$
: ExR_j \rightarrow \left(\forall \sum_{i=1}^{n} ans_i \in Q_i \right) \otimes f(Cof)
$$

(Cof) symbolizes the Confidence Value. We have employed multiple answer notions associated with each fact. Thus, we have to choose the best answer among all the provided answers. In "CogQA_Edu" multiple answers are extracted from the documents (which have been provided by the group of teachers) using (E, A, OS) extraction method. Now there is a question: how to choose the best answer among all the available answers? For that we have a group of experts (five experts are there). They grade or rank each answer on [0,..,10] point scale. Subsequently, "CogQA_Edu" takes the mean of all the grade values and selects the highest value as the best answer for the question. (AnsCof) is a function, which associates Confidence Value with each answer of the fact (a particular fact containing multiple answers). Subsequently, we select the highest Confidence Value answer as the best answer to the question. This function is used for searching the best answer from a list of provided answers. "QAEdu: CIT" also offers answer suggestion in addition to the best answer.

(d) Define "Suggested Answer" for finding the "Answer Suggestion".

$$f(AS) := \exists, (ans_i, Q_i) : [ans_i \in Q_i] \ \&\& \ (\forall, (E,A) \to \exists, AS)$$
$$(ans_i \leftarrow \{KB \otimes f(L)\}) : \# |ans_i|_{largest} \quad (10.26)$$

f(AS) is a function provided by the "CLVM" engine, which endows with a best cardinality answer as the "Suggested answer" (recommended by "CogQA_Edu") to each question based on previous Liking Values of users. This uses (AnsL) function for answer suggestion.

(e) Define "Best Answer" for finding the suggested answer.

$$f(AB) := \exists, (ans_i, Q_i) : [ans_i \in Q_i] \ \&\& \ (\forall, (E,A) \to \exists, AB);$$
$$\therefore (ans_i \leftarrow \{KB \otimes f(Cof)\} \uplus QM_f) : \# |ans_i|_{largest} \quad (10.27)$$

f(AB) is a function provided by the "CLVM" engine, which provides the best cardinality answer as the "Best answer" (among available answers) to each question based on Confidence Values (supplied by the group of Experts). This uses (AnsCof) function for selecting the best answer.

(f) Define (QM) Query Module Function for dealing with user questions.

$$QM_f := \exists, Q_i : [((x_i, y_{i.}) \ni LCom) \leftarrow Q_i]$$
$$\&\& \ \left(\exists, ans_i \in \sum_{i=1}^{n} ans_i \in Q_i\right) \otimes ((Cof) \in AnsCof) \quad (10.28)$$
$$: \left[\# |ans_i|_{largest} \leftarrow (agent(KB_k) \otimes (x_i, y_{i.}))\right] \leftarrow (DM_f)$$

The query modeling function extracts the best answer from the knowledge base by using the CLVM engine's "Confidence Value Function". It uses the DM module during this process.

(g) Define (DM) Decision-Making Module for helping the QM module in the process of answer extraction.

$$DM_f := (DM \in QM) \ \&\& \ [DM \ni (RM \uplus I \uplus Rul's \uplus NLP_f)];$$
$$\therefore \left(\exists,; ans_i : (\# |ans_i|_{largest}) \in X \leftarrow (KB),; \ \&\& \ X \equiv \sum_{i=1}^{n} ans_i\right) \quad (10.29)$$

The DM module is an element of QM module. It internally consists of the following sub-modules: the "Reasoner" (RM), the "Inferencing" (Inf), set of "Rules", and the "Natural Language Processing Functions" (NLP_f) modules.

(h) Reasoner Module (RM) consists of the following two tuples {NCM, KC}:

$$\begin{cases} NCM := NLP \ Question \ Corpus \ Module. \\ KC := Knowledge \ Corpus. \end{cases} \qquad (10.30)$$

10.3.2 QUESTION UNDERSTANDING

Figure 10.2 explicates how a particular question is analyzed by a NLP interface. A student asks the question "Q" and NLP functions analyze it. NLP offers the power of understanding the semantics of the question. NLP functions (tokenization, POS tagging, Chunking, EAR, Algorithms) use the NLP Q-Corpus for this; subsequently, we fill the question template. Now we explore the knowledge corpus with the help of confidence and Liking Value function for finding the missing values (the Best Answer (using Confidence Value function), Suggested answer (using Liking Value function)) of the question template. The central element of a Q-A system is the semantical and syntactical categorization of the user questions asked in the natural language form.

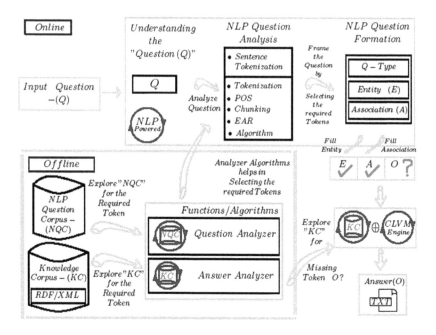

FIGURE 10.2 Schematic diagram for illuminating how a particular question is analyzed by a (NLP interface). A student asks the question "Q" and (NLP-functions: tokenization, POS - tagging, Chunking, EAR, Algorithms) analyzes it for understanding the semantics of the question. Subsequently, we fill the question template and discover the Knowledge Corpus using the above template.

If the Q-A system can precisely comprehend the semantics of the question, it can perform better answer searching and extraction. So the crucial step is to classify the question domain and the type of question. There are two types of Q-A systems [59] based on application domain (General Domain, Restricted Domain [60]). General domain questions deal with the open questions. Here, the excellence of the answers has always negotiated due to the broad domain. In QAEdu, we are dealing with the questions of "Electrical Motors and Basic Electromagnetic Concepts". Thus, the focus of the question should be the elementary electrical motor and electromagnetic concepts. The second step is identifying the type of question and linguistic components present in the question. Here, Qtype facilitates in selecting the appropriate answer XML file in the knowledge corpus as QA_Edu implements separate knowledge corpus for each Qtype category. The interrogative pronouns itself describe the kind of questions; they have some inherent meaning, for example,

$$1. \ Who \xrightarrow{Represents} (Person/Entity).$$

$$2. \ When \xrightarrow{Represents} (Date/Time).$$

$$3. \ What + Verb \xrightarrow{Represents} (Date/Time).$$

$$4. \ What + Noun \xrightarrow{Represents} (Person/Entity).$$

Thus, based on question nature, we can categorize questions in the following forms [61]:

(1) Factoid type questions (What, When, Which, Who, How)
(2) List type questions
(3) Causal questions (Why; How)
(4) Boolean Questions (Yes, No)

In QAEdu we are presently covenanting with factoid type of questions (Shortly, we will try to put into practice the other category of questions). At this point, we are using NLP interface for identifying, understanding, and extracting the components of questions.

Algorithm 10.1 explains the logical description of the extraction process of ("WP – Pronoun") (Question-type) from the user questions. Firstly, we group the individual tokens of the sentence in the group of "Noun Phrases", and "Verb Phrases" by using the basic "NLP" functions ("Tokenization", "POS Tagging", and "Chunking"). Subsequently, we use a "Hybrid Model" (a combination of ML + Rule-based techniques) to extract the required "WP-Pronoun" from the user questions. Tables 10.1 and 10.2 we have recorded the performance of the "Hybrid Model" and found the satisfactory results.

After extracting the "WP-pronoun", we search for the required "Q-type" mentioned in the "Q-type" list defined under the "NLPQ-Corpus". If the match is successful, the system declares the positive result otherwise an "error" message is displayed. This "Q-type" extraction process plays an important role while extracting the best answer from the knowledge base because the knowledge base is further subdivided in the small knowledge bases headed by a particular "WP-Pronoun".

TABLE 10.1

The "Question-Type" Classification for "CogQA_Edu" System. It Utilizes Machine Learning Classifiers and Rule-Based Classifier. Subsequently, It Found That the Rule-Based Classifier Works More Accurately in Comparison to Machine Learning Classifiers

Algorithms	Count Vectors	Word-Level TF-IDF Vectors	N-gram Level TF-IDF	Character Level TF-IDF
Linear Classifier (Logistic Regression)	0.96	0.92	0.75	0.92
Naïve Bayes (NB)	0.96	0.78	0.73	0.74
Random Forest (RF)	Word-Level "TF-IDF" Vectors Count Vector			0.960.92
SVM Classifier	N-Gram Level "TF-IDF" Vectors			0.72
Rule-Based Classifier	Regular Expression & Rules			0.99

TABLE 10.2

The Table Explains the Semantic Components (Entity and Association) Extraction Analysis Results for Questions in "CogQA_Edu"

Method	P (Precision)	R (Recall)	F (F-Measure)
Rule-based classifier approach	95.20	97.20	96.18

10.3.2.1 Evaluation of Extraction Process for Qtype and Its Semantic Components

Table 10.1 denotes the result of the assessment of Q-type component extraction from the user questions. We have applied two approaches: Machine Learning and a rule-based approach. For that we have collected sample 500 questions inspired by the TREC - QA-(9,10,11) questions set (TREC - QA 9; 10; 11 contains 693; 500; 500 questions, respectively). The selected questions are of the following category (What, When, Why, Where, How, Who). Afterward, we developed some Machine Learning classifiers like Naive Bayes, Random Forest, Support Vector Machine and tested the selected sample of questions by the help of these classifiers. We have also designed a rule-based classifier with the help of Regular Grammar and Set of Rules and found that the accuracy of rule-based classifier is the highest (which is 0.99) in contrast to Machine Learning classifiers (which is 0.96 with Naive Bayes and Logistic Regression, when used with Count Vector method). Thus we have applied rule-based classifier subsequently for the extraction of Qtype component from the user questions. After getting outstanding results from the Rule-Based classifier, we have applied the same method in the extraction of semantic components (E, A) from the user question. Table 10.2 denotes the result of the assessment of semantic components (E, A) extraction from the user questions, and we got an excellent precision that is (95.20), the Recall is also very high (97.20), and the corresponding F-Measure score is (96.18). (95.20), the Recall is also very high (97.20), and the corresponding F-Measure score is (96.18).

ALGORITHM 10.1   Q-TYPE EXTRACTION FROM THE USER QUESTIONS

Procedure:

Query (Q, Q-type): (I/P:- Q ∈ Question & O/P:- Q-type)
   A_Com [n] ← Perform (Tokenization(Q))|-[({x_i, y_i.} ≡ Com) ∈ Q]
   (P_Com[n] ← Perform (POS (A_Com)))
   (CP_Com[m] ← Chunking (P_Com))
   for (i=0) to (CP_Com.length) do
      Chunk1 ← Extract_Qtype_HModel('chunk')
   end for
   if('chunk1' ∈ 'Q_type_list') then
      Fill_Q-template('Q_type') : 'Q_type' -| ⟨Q_type,E,A⟩
      Set (Flag = '1')
   end if
   if (Flag=='1') then
      Message: (Q-type found successfully)
      Call: procedure Classify_E_A(Q,(E,A))
   else:
      Message: (Reframe the Question)
   end if
end procedure;

10.3.3 Linguistic Component Extraction (Entity Disambiguation and Association Detection) from the Question

After the classification of the questions, the next step is the extraction of linguistic components from the input natural language query. We try to extract linguistic components and fill the question template accordingly. This template takes the following form (QType; Entity; Association; Object). It pursue (Subject; Predicate; Object) in the sentence structure and rest tokens are discarded. At this juncture, we only concerned for the QType; Entity; Association pair, as the object token is missing presently. The output of this phase supplied as an input for the configuration of the query. Following (Examples 10.3 and 10.4 and Algorithm 10.1) elucidates the extraction process of linguistic components (QType; Entity; Association) from the user query.

- **Attributive Searching of linguistic Components:**
  The accomplishment of the valid tokens is essential and significant while analyzing the input NL question. During the linguistic analysis of the questions, we have to take care of the following points. As we know, that the natural languages have many ambiguous words as well as there are some additional words also like adjectives and adverbs (which qualifies their associated words) that can change the meaning of a sentence.

EXAMPLE 10.1

> Question: What is the capital of India?
>> (India, Capital).
> Answer: Delhi.

If the user modifies the question to "What is the {commercial capital} of India?" Therefore, the adjective (Commercial) definitely will amend the meaning of the entire sentence, and the answer will also need to adjust accordingly.

EXAMPLE 10.2

> Question: What is the commercial capital of India?
>> (India, Commercial Capital)
> Answer: Mumbai.

Thus by taking into account the points mentioned above, in QAEdu we are using Attributive and Dictionary-based searching of linguistic components. POS and Chunking (NLP Functions) help in the implementation of such robust and exclusive searching methods (refer to Examples 10.3 and 10.4). Part of speech tagging (POS) tags the corresponding POS tag with each token of the sentence so that we can promptly identify which particular word is adjective/adverb/noun, etc.

Subsequently, the Chunker groups the tokens into a set of NP (Noun Phrase) and VP (Verb Phrase) groups. It is the most elegant way to correlate adjectives with their corresponding nouns. By applying some rules and with the help of chunker output we can perform attributive searching of the tokens very efficiently. It increases the question understanding power of QAEdu refer (Examples 10.1 and 10.2).

EXAMPLE 10.3

O/P of "POS" function.

$$\begin{bmatrix} What \, / \, WP \ \ is \, / \, VBZ \ \ the \, / \, DT \ \ commercial \, / \, JJ \\ capital \, / \, NN \ \ of \, / \, IN \ \ India \, / \, NNP?/. \end{bmatrix}$$

(Example 10.3) shows the yield of the "Parts of Speech" (POS) function offered by NLP interface. Here, we can label each token of the sentence by its corresponding (Part of Speech) tag. What (WP: - WH pronoun), is (VBZ: - Verb), the (DT: - Determiner), commercial (JJ: - adjective), capital (NN: - Noun), of (IN: - Proposition), India (NNP: - Proper Noun).

EXAMPLE 10.4

O/P of "Chunker" Function.

$$
\begin{bmatrix}
\begin{bmatrix} NP\ What\ /\ WP \end{bmatrix} \\
\begin{bmatrix} VP\ is\ /\ VBZ \end{bmatrix} \\
\begin{bmatrix} NP\ the\ /\ DT\ commercial\ /\ JJ \\ \quad capital\ /\ NN \end{bmatrix} \\
\begin{bmatrix} PP\ of\ /\ IN \end{bmatrix} \\
\begin{bmatrix} NP\ India\ /\ NNP \end{bmatrix} \\
\begin{bmatrix} ?/\ . \end{bmatrix}
\end{bmatrix}
$$

Example 10.4 shows the yield of the Chunker module. It is apparent that it solves the problem of attributive searching as we are merely interested in the NP (Noun Phrase: - [NP the/DT commercial/JJ capital/NN]). So merely by applying some rules and functions, we can solve our purpose.

Algorithm 10.2 provides the logical explanation of the linguistic components extraction process present in the user's questions. For searching the best answer from the knowledge base we have prepared a next level of question template (Q-type; E; A). Algorithm 10.1 explains the extraction process of (Q-type), and the other components are extracted using Algorithm 10.2. Thus, by using the "NLP Functions", we firstly convert the tokens into the required form. Subsequently, we have developed another model (a Rule-Based Model) for extracting the other tokens from the user question. The reason behind choosing this model is the performance of the model. We have got the satisfactory performance by using this model (the performance results are mentioned in Table 10.2). If the model is able to successfully extract the required tokens from the questions, it shows a positive message; otherwise it shows an error message.

- **Dictionary-based searching of linguistic components**
  There are several words in the natural languages (like English) which have identical meaning. We humans can recognize the resemblance and disparity between such words, simply by learning these words. However, for a machine, it is an exigent task to discriminate among these words. Therefore, for escalating the clout of QAEdu, we have employed the Dictionary concept. For example, (Q: What is the definition of induction motor?). So, the triplet-based searching should be (What, induction motor, definition).

  However, the user may replace the word "definition" by the following words ("Mean by", "Stand for", "Principle", "Think of", etc.) and can reframe the question which will have the same significance. Thus, we can define all these words as the synonyms of the particular word "Definition" and can place them in the dictionary. This functionality augments the question understanding power of the QAEdu(refer to Figure 10.8).

ALGORITHM 10.2 ENTITY AND ASSOCIATION DETECTION

Procedure:

$$Classify\_E\_A\big(Q,(E,A)\big):\begin{pmatrix} I/P:-Q \in Question\ \&\& \\ O/P:-E \in Entity\ Set,\ A \in Association\ Set \end{pmatrix}$$

$A\_Com\ [n] \leftarrow Perform(Tokenization(Q))|-[(\{x_i, y_i, .\} \equiv Com) \in Q]$
$(P\_Com\ [n] \leftarrow Perform(POS(A\_Com)))$
$(CP\_Com[m] \leftarrow Chunking(P\_Com))$
$for(i = 0)\ to\ (CP\_Com.\ length)\ do$
  $(chunk1,\ chunk2) \leftarrow Extract\_E\_A\_RModel(`chunk')$
$end\ for$
$if(`chunk1' \in `E\_List')\ then$
  $Fill\_Q - template(`E') : [E-|\langle Q\_type, E, A\rangle]$
  $Set(Flag = `1')$
$end\ if$
$if(`chunk2' \in `A\_List')\ then$
  $Fill\_Q - template(`A') : [A-|\langle Q\_type, E, A\rangle]$
  $Set(Flag1 = `1')$
$end\ if$
$if(Flag = \ = `1')\ then$
  $if (Flag1 = \ = `1')\ then$
    $Message : (E, A\ Pair\ is\ successfully\ found).$
  $else:$
    $Message : ("Association"\ value\ is\ missing\ in\ the\ question).$
  $end\ if,$
$else:$
  $Message : ("Entity"\ value\ is\ missing\ in\ the\ question).$
$end\ if,$
$end\ procedure;$

10.4 IMPLEMENTATION OF EXPERIENCE AND KNOWLEDGE MODEL

10.4.1 KNOWLEDGE-BASE CONSTRUCTION

For any particular Q-A system, knowledge base is the main component. Thus the vigorous knowledge base is mandatory for a good Q-A system. In the QAEdu system, we are encouraging the experiential Knowledge Base concept to overcome the problems of accuracy and genuineness of the answers. Here, we have a group of teachers who provide the required documents for knowledge construction and also work as an expert for the assessment of the system. In QAEdu teachers offer the set of documents, and the classifier is going to mine the documents for extracting (E; A; OS) (Entity, Association, Object-Set) triplets. The (E; A; OS) triple symbol-izes a fact, and the collection of triples are Facts. It is a sort of automatic process (mining of the triples from the documents and accumulating them in the knowledge base); for that, we have developed a rule-based classifier (Due to the high accuracy of this classifier, we selected this category of classifier.). If a teacher has a better answer for a particular question, he can directly fill the answers in the knowledge base. Subsequently, for practically representing the triples in the knowledge base, we are following the XML syntax structure. In the knowledge base, we have separate XML file for each question type (Like for "What" category questions we have a sepa-rate XML file and for others also we have a different file). Under each Qtype, we have a pair of (Entity, Association) and with each (E, A) pair many answers (object set) are associated. Now there arises a question of choosing the best answer among all the available answers; thus the system QAEdu uses the CLVM engine for serving the essential purpose. CLVM engine uses the notion of Confidence and Liking Value for providing the best answer and suggested answer, respectively, refer to Section 10.4.2 and Figures 10.3–10.6 for further elucidation. Figures 10.3 and 10.5) repre-sent the functioning of Confidence and Liking Value utilities and figure (Figures 10.4 and 10.6) denotes the practical implementation of these functions. Since QAEdu is a cognitive system; thus the appraisal of the system should be done by a human expert rather than the traditional methods. Section 10.4.1.1 explains the evaluation of extraction process of triples from the documents.

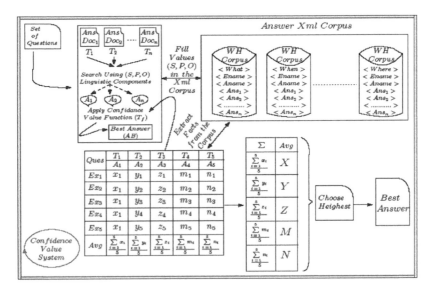

FIGURE 10.3   Schematic diagram showing the process of "The Best Answer" extraction by the use of "Confidence Value Function".

```
<?xml version="1.0" encoding="utf-8" standalone="yes"?>
<!--This is an Expert Grading XML File for fetching the best answer from t
<ExpertGradingCorpus Qtype="what">
  <GradeAnswers GID="0">
    <EName>Electric Motor</EntityName>
    <AName>Definition</AssociationName>
    <Answer>Electric motors operate on three different physical principles
    <Values>
      <ExpertMarks>
        <Value>7</Value>
      </ExpertMarks>
      <ExpertMarks>
        <Value>8</Value>
      </ExpertMarks>
      <ExpertMarks>
        <Value>8</Value>
      </ExpertMarks>
      <ExpertMarks>
```

FIGURE 10.4 The snapshot is representing the list of Expert Values for the implementation of the Best Answer.

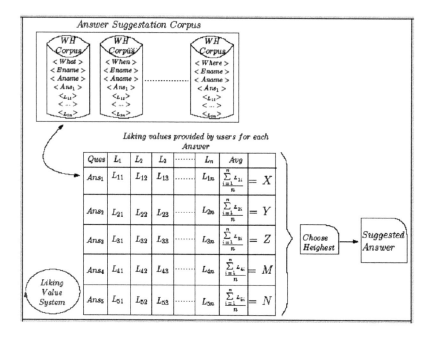

FIGURE 10.5 The schematic diagram is showing the process of "Suggested Answer" extraction by the use of "Liking, Value Function". In addition to "The Best Answer" "CogQA_ Edu" provides "Answer Suggestion" to the users based on the "Highest Liking Answer Concept".

```
 AnswerGradingForSuggestedAnswer.xml - Notepad                              - □ x
File Edit Format View Help
<?xml version="1.0" encoding="utf-8" standalone="yes"?>
<!--This is an Answer Grading XML File for performing various stastical op
<AnswerGradingCorpus Qtype="what">
  <GradeAnswers GID="0">
    <EName>Electric Motor</EntityName>
    <AName>Definition</AssociationName>
    <Answer>Electric motors operate on three different physical principles
    <Values>
      <LikeValue>
        <Value>9</Value>
      </LikeValue>
      <LikeValue>
        <Value>9</Value>
      </LikeValue>
      <LikeValue>
        <Value>7</Value>
      </LikeValue>
      <LikeValue>
```

FIGURE 10.6 The snapshot is representing the list of Liking Values for the implementation of Suggested Answer. In addition to the "Best Answer" "CogQA_Edu" implements the concept of "Answer Suggestion" by the system.

10.4.1.1 Evaluation of Extraction Process of Triples from the Documents for the Construction of Knowledge Base

This section elucidates the assessment process of triples extraction from the documents provided by the teachers (domain experts). The teachers have provided the documents according to the following category of questions (hasType, hasFunction/Property, hasComponents, hasFact, hasApplication). Thus Table 10.3 represents the Precision (P), Recall (R), and F-measure (F) scores accordingly. The classifier extracts the triples from the documents in the (S; P; O) (Subject, Predicate, and Object) format. Further we are going to store them in the (E; A; OS) format (here (E) is akin to (S), (A) resembles (P), and (OS) is similar to (O)) in the knowledge base. Except that for each (E; A) pair, we have a set of answers; thus (OS) symbolizes "set of answers". We have a total of 250 documents for different categories (for hasType we have 58 documents, for hasFunction=Property we have 61 documents, for has-Component category we have 56 documents, for hasFact category we have 42 documents, and for hasApplication category we have 33 documents) provided by the teachers.

TABLE 10.3

"EAO" Triple Extraction Based on Different Categories (hasType, hasFunction, hasComponent, hasFact, hasApplication) for "CogQA_Edu".

Analysis/Sentence Categories	hasType	hasFunction/ Property	hasComponent	hasFact	has Application
P (Precision)	94.82	95.08	92.85	95.23	100
(Recall)	100	96.72	98.21	100	100
F (F-Measure)	97.34	95.89	95.45	97.56	100

TABLE 10.4
Overall Analysis for "EAO" Triple Extraction for "CogQA_Edu".

Analysis/Sentence Categories	P (Precision)	R (Recall)	F (F-Measure)	Overall Accuracy
Rule-Based Approach	95.20	98.8	96.96	95.20

Subsequently, Table 10.4 represents the over overall accuracy (P; R; F) of the triples extraction process. Here, the overall precision value for all category sentences is (95.20), the recall is (98.8), and the F-Measure score is (96.96). This classifier shows that the overall accuracy for (S, P, O) extraction is (95.20), which is a good score.

10.4.2 METHODS FOR ANSWER SEARCHING AND EXTRACTION FROM THE KNOWLEDGE BASE

Based on the information representation systems, we have to use particular techniques for answer mining from such stores. These days, we are using three kinds of knowledge representation techniques [59], structured, semi-structured, and unstructured. Structured knowledge uses tabular format for storing data, for example, SQL, MySql, SQLite, DB2 offer a structured representation of information. In semi-structured (like XML) representation we use tags. Tag works as metadata, advances the user understanding, and is appropriate for online data exchange systems. In unstructured representation, we deal with textual data; it has no format for representing data. These systems use NLP and IR techniques for mining of data. Thus for each representation, unique techniques have been developed for extracting the information from these data stores. We can classify the Q-A systems based on the approaches used for answer extraction. Basically, there are two types of approaches for answer extraction Web mining-based answer extraction [62] and knowledge-based answer extraction [63, 64]. To serve the purpose of building QAEdu we are using experiential knowledge base, in which knowledge is represented (E, A, O) triples. For this, we have collected a set of documents from the group of teachers (of Electrical Engineering domain).

Here, we are following multiple answers approach for a single fact (as we are collecting documents from a group of teachers and each teacher has its own opinion on a particular topic. That's why the system has a set of answers for each fact). For the justification of answer extraction approach, we are scoring the answers based on (Confidence Value and Liking Value function) criteria. At present in QAEdu we are using Confidence Value for selecting the best answer among the offered answers and the Liking Value function for the answer suggestion.

- (Confidence Value System): The proposed architecture uses the Knowledge Base approach for answer extraction. We are using the experience-based multiple answers-based knowledge base approach. We have multiple answers (provided by a group of teachers). Therefore, to find out the best answer among them, we have consulted a group of experts (5 Experts panel) they are going to give scores for each answer extracted from the documents provided by the teachers on [0,...,10] scale. Further, the system is going to find out the average score of each answer, the highest average score answer is the winner and will be picked up for best answer (refer to Figures 10.3 and 10.4).

- (Liking Value System): Liking Value function helps in answer suggestion and implements the knowledge-based concept also. As the system interacts with the users, it maintains the Liking Value matrix provided by the users for each answer (in XML format). When all the answers are marked using the Liking Value function, it follows the knowledge base approach. According to the best Liking Value, it selects the suggested answer. At this point, it takes the average of all the Liking Values associated with a particular answer (different users may grade a specific answer by different grades according to their understanding). Subsequently, we compare each answer, according to their average Liking Value, and select the highest average value answer (which will be the best answer suggestion). However, in the case of a tie (refer Example 10.5) (when two answers have the same average Liking Value), it further checks which answer has got the highest Liking Value among all values associated with each answer and select that particular answer and resolves the tie refer (Figures 10.5 and 10.6). Following (Example 10.5) explains the concept of the conflict resolution process, which can arise while selecting the best Liking Value when two answers have the same average value scores.

EXAMPLE 10.5 CONFLICT RESOLUTION IN THE CASE OF CONFLICT WHILE EXTRACTING THE BEST VALUE

If we have two answers (A_1) and (A_2) having Liking Value as follows:

$$A_1 := (5,6,7,5,6,7,5,6,5,5)$$
$$A_2 := (9,7,5,6,7,5,6,7,2,3)$$

- Claculate the average value for each answer:

$$\text{Average}(A_1) = 5.7;$$
$$\text{Average}(A_2) = 5.7;$$

- This is the case of a tie situation:

$$\left(\begin{array}{l} \text{Select the highest liking value answer as the} \\ \text{best answer} \end{array} \right).$$

Therefore select (A_2)

10.5   METHODOLOGY

This section of the chapter covers the practical implementation features of QAEdu, an NLP-based Q-A system, by using an appropriate flowchart representation.

Above flowchart (refer to Figure 10.7) explains the construction process of the following modules, Concept Model (NCM), Experience and Knowledge Model

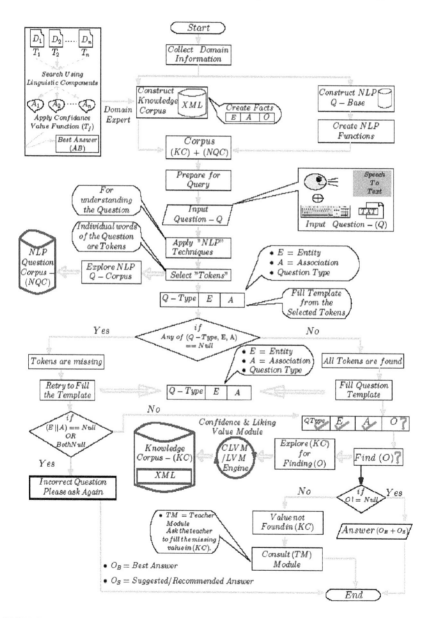

FIGURE 10.7   This flowchart is showing the step-by-step construction process of an NLP-based Cognitive Question-Answering system ("CogQA_Edu"). It explicates the construction process of the following modules, Concept Model (NCM), Experience and Knowledge Model (KCM), and the Interface model (UIM).

(KCM), and the Interface model (UIM). First of all, we accumulate the domain information, delineate the facts, and fill them in the knowledge base. At this instant, the system is ready to handle the user's queries. A user asks the query and QAEdu explores its knowledge base using the CLVM engine and finds out the Best, and the Suggested Answer. The domain information has stored in the knowledge base in the form of triples using XML file representation. Each triple (E; A; OS) represents a Fact and the collection of Fact constitute the Facts or Knowledge. We have a list of documents provided by the teachers (Domain Experts). The triples are extracted from these documents using the classifier (knowledge engineer helps in this process). We are using a triplet format (Entity, Association, Object Set) for representing the fact (the object set refers the group of answers as we have a list of answers for each (E; A) pair). Subsequently, we prepare NLP Q-Corpus also that helps NLP functions to classify the correct semantics of the question. At this moment both the corpus are ready; thus we can go for the querying phase; the query may be asked either by using a text processing module or using a speech recognition module (proposed). At this instant using NLP functions, we comprehend the question and try to fill (Q - Type, E, A) question template. Now, we check whether all the tokens of question template have been filled successfully or not (If there is any missing token in the template, we declare that the question is not framed appropriately and ask the user for reframing the question). If all the tokens are found, we use the template for searching and extracting the answers (Best answer, Suggested Answer) from the knowledge base. Afterward, we explore the knowledge base using RM-Module. Here the concept of confidence and Liking Value has been used. We use the Confidence Value function for getting the best answer from the KB and the Liking Value function for the suggested answer. If the value has been found in the KB, bring it and return to the user (as an answer). Otherwise, return false and ask the teacher TM-Module to provide the required documentation. QAEdu also supports direct insertion of facts in the knowledge base if the teacher found it suitable.

10.6   IMPLEMENTATION OF INTERFACE MODEL

10.6.1   EXPERIMENTAL DESIGN AND RESULT

In this section, we try to demonstrate the practical implementation of all the modules of QAEdu. We have chosen Visual-Studio IDE and C# language for the execution of this Q-A system and for creation of knowledge-corpus, the XML syntax has been used. For representing facts that consist of long answers, ontology is not a feasible approach. Thus, we have chosen XML syntax for storing the facts in the knowledge base [13, 15]. For knowledge mining, we are using LinQ querying language. Here, we have implemented the full power of NLP provided by SharpNLP (open source, English language natural language processing tools written in C#). It is an English language natural language processing tools written in C#. Centered around a port of the Java OpenNLP library (http://opennlp.sourceforge.net/), which includes a sentence splitter, a tokenizer, a part-of-speech tagger, a chunker (used to "find non-recursive syntactic annotations such as noun phrase chunks"), a parser, a name finder, and a coreference tool. For the validation of QAEdu system Python language has been chosen which provides the smooth implementation of Machine Learning concepts and regular expressions.

10.6.2   Interface Module Designing

QAEdu primarily have five components at the interface level.

(1) TM is accountable for accumulation and updation of facts (triples) in the knowledge corpus (KC). Contrasting rule-based systems QAEdu as an autonomous system, and for attaining the cognitive capability, it supports updating and amendment of new facts as required.

(2) NLP Q-Corpus-Module is accountable for supervision of question corpus. A clear understanding of the question is the crucial part of any Q-A system as the semantics of the question helps in finding the best decision, and it also decides the performance and credibility of the system. NLP (Natural language processing) functions provide this capability to QAEdu.NLP functions and algorithms use the question corpus for understanding the user query(refer to Figures 10.8, 10.9, and 10.11 for Dictionary Concept, Domain Identification Concept, and QAEdu's NLP Corpus Implementation, respectively).

(3) KC knowledge corpus is liable to manage knowledge base. Knowledge is stored in the form of triples (E; A; O). It has assumed that a robust knowledge base is worth more than a good quality algorithm. Thus to achieve the cognitive computing capability one should use a robust knowledge base. QAEdu uses XML syntax for modeling the triples, and the set of triples forms up the knowledge base. The teacher provides necessary documentation for the knowledge extraction and may store the facts directly in the knowledge base if required. He/she can also modify the existing triples and can add a new one as needed. Refer to Figures 10.10, 10.6 and 10.4 for XML representation of Facts (E, A, O), Answer Grading Corpus for fetching the suggested answer from knowledge base, and Answer Grading Corpus for fetching the best answer from knowledge base, respectively.

```
DictionaryValuesXMLFile.xml - Notepad
File  Edit  Format  View  Help
<?xml version="1.0" encoding="utf-8" standalone="yes"?>
<!--This is a Dictionay XML file for increasing Question understanding capability-->
<Dictionary>
  <Element ID="1">
    <AName>Definition</AName>
    <SynonymValues>
      <Synonym>
        <SynonymName>Think</SynonymName>
      </Synonym>
      <Synonym>
        <SynonymName>Understand</SynonymName>
      </Synonym>
      <Synonym>
        <SynonymName>Mean</SynonymName>
      </Synonym>
      <Synonym>
        <SynonymName>Stand</SynonymName>
      </Synonym>
    </SynonymValues>
  </Element>
```

FIGURE 10.8 This snapshot represents the implementation of dictionary concept (for providing the synonym understanding capability to the "CogQA_Edu" system).

```
DomainIdentificationEntityList.xml - Notepad
File  Edit  Format  View  Help
<?xml version="1.0" encoding="utf-8" standalone="yes"?>
<!--This file contains possible list of words which belongs to the Electric Motors,
<DomainIdentificationCorpus>
  <Entites ID="0">
    <EntityName>AC Motor</EntityName>
  </Entites>
  <Entites ID="1">
    <EntityName>DC Motor</EntityName>
  </Entites>
  <Entites ID="2">
    <EntityName>Induction Motor</EntityName>
  </Entites>
  <Entites ID="3">
    <EntityName>Synchronous Motor</EntityName>
  </Entites>
  <Entites ID="4">
    <EntityName>Slip</EntityName>
  </Entites>
  <Entites ID="5">
```

FIGURE 10.9 This snapshot represents the List of Entities for the implementation of Domain Identification concept. "CogQA_Edu" can answer the questions related to "Electrical Motors" and "Basic Electromagnetic".

```
FactsCorpus.xml - Notepad
File  Edit  Format  View  Help
<?xml version="1.0" encoding="utf-8" standalone="yes"?>
<!--This is an Xml File for Facts Corpus (Answer Corpus)-->
<Facts Qtype="what">
  <Fact Fact_ID="0">
    <Ename>Electric Motor</Ename>
    <Aname>Definition</Aname>
    <Answers>
      <Answer_Info A_ID="0">
        <AnswerValue>Electric motors operate on three different physical principles
      </Answer_Info>
      <Answer_Info A_ID="1">
        <AnswerValue>An electric motor is an electric machine that converts electri
      </Answer_Info>
      <Answer_Info A_ID="2">
        <AnswerValue>a device that converts electrical energy to mechanical torque.
      </Answer_Info>
      <Answer_Info A_ID="3">
        <AnswerValue>An electric motor is a device for converting electrical energy
      </Answer_Info>
```

FIGURE 10.10 This snapshot represents the list of "Facts" for the implementation of "Knowledge Base".

(4) QM Query module handles the user queries. It implicitly uses RM module and CLVM engine for inferencing the knowledge base. RM module explores the KB and uses the concept of inheritance for establishing relationships between entities.

(5) A Cognitive Q-A System QAEdu: refer Figure 10.12, represents the interface for the QAEdu system.

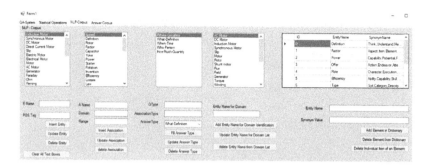

FIGURE 10.11   This screenshot represents the NLP Q-Corpus. It represents the implementation of "Concept Model". It has list of "Entities", "Associations", "Q-type", "Domain Identification Entity List", and the "Dictionary Concept". Using this corpus, "NLP" functions analyze the user questions and try to extract "Q -Type" and "Semantic Components" and fill them in the "Q-Template (Q-Type, E, A)". Subsequently, this template is used to search for answers from the knowledge-base.

10.6.3   RESULT

Figure (refer Figure 10.12) illustrates the practical implementation of QAEdu Q-A interface. A user query "What is the definition of an electrical motor" has answered in this module (all the functions are carried out in the background, to achieve the abstraction). Subsequently, NLP functions utilize NLP Q-corpus for understanding and extracting the linguistic components (At this point, Q-Type = "What", Entity = "Electrical Motor", Association = "Definition") from the query. After getting the necessary components, QAEdu explores the knowledge base using query module QM which implicitly uses the RM and CLVM module and provides the answer. Afterward, we are using a grading scale of (−10; .. 0 .. ; 10) so that the user can grade the answer according to his/her liking. Each answer maintains a Liking Value matrix and uses it in the process of answer suggestion. Second group box shows how QAEdu provides the suggested answer (visual analysis component explains how it takes the average of all the Liking Values associated with each answer and selects the best value and its associated answer as a suggested answer).

10.6.4   EVALUATION OF QAEDU QUESTION-ANSWERING SYSTEM

This section of the chapter explains the evaluation process of QAEdu. We framed the sample (30) questions based on TREC-QA-9, 10, 11 questions set (TREC - QA 9, 10 11 contains 693, 500, 500 questions, respectively) for the assessment of QAEdu system. The selected questions are of the following categories (What, When, Why, Where, How, Who). Traditionally, in IR systems there are following parameters for evaluating the system accuracy, Precision (P), Recall (R), and F-Measure. But QAEdu is a Hybrid Cognitive Architecture inspired Cognitive Q-A system. And Cognitive Systems aim to mimic humans so there should be some other criteria in which the accuracy and efficiency should be measured by some human expert. Thus for such Q-A systems, there are following parameters for performance evaluation (TREC-QA,

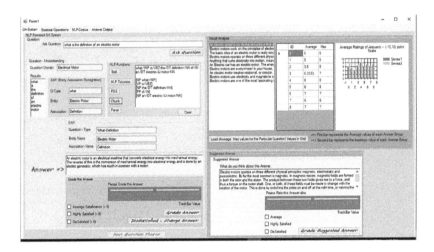

FIGURE 10.12 This screenshot represents the "Q-A Interface" of "QAEdu: CIT" an "NLP"-based "Cognitive Question-Answering System". It consists of three modules "Question", "Suggested Answer", and "Visual Analysis" module. In the "Question" module a user asks the question related to "Electric Motor and Electromagnetic Domain". This module extracts the "Q-Type" and "Linguistic Components" from the question and explores the "Knowledge-Base" using the "CVM" engine and "RM" module for the extraction of the "Best Answer". In the second module "Suggested Answer" the "CogQA_Edu" system extracts the suggested answer to the user using the "CLVM" engine. The third module "Visual Analysis" shows the various statistical operations involved in the answer extraction process. Also according to the liking of users about the answers users are going to give grades to the answers.

Performance Analysis). The evaluation metrics should consists of following parameters, First Hit Success (FHS), First Answer Reciprocal Word Rank (FARWR), First Answer Reciprocal Rank (FARR), Total Reciprocal Word Rank (TRWR), Total Reciprocal Rank (TRR), Mean Rank Reciprocal (MRR), and Precision (P).

- First Hit Success (FHS) if the first answer provided by the QAEdu is correct then the FHS value is (1), otherwise (0).
- First Answer Reciprocal Rank (FARR): if the first answer provided by the QAEdu is correct then FARR is (1/1) = (1). If the fifth answer is correct, then FARR is (1/5) = (0.20). If any of the responses provided by the system is not correct, then the value is (0).
- First Answer Reciprocal Word Rank (FARWR): Signifies the distance of appropriate word in the answer provided by the system. Let, the 9th word from the starting is the proper word then the value will be (1/9 = 0.11).
- Total Reciprocal Rank (TRR): QAEdu supports multiple answers for each (E, A) pair then it might be possible that there may exist more than one correct answers. Assume if the first and fourth answer is correct then (1/1 + 1/4) will be the final value.
- Mean Reciprocal Rank (MRR): calculates the average over a set of n queries.
- Precision (P): Precision is the fraction of retrieved documents that are relevant to the query.

Table 10.5 shows the evaluation results of the question understanding power of QAEdu. Human experts checked the QAEdu system's performance on the parameters as mentioned earlier and found that the FHS rate is (97%), FARR rate is (97%), and the MRR is also very high (0.9). This is because of the experiential knowledge base and Confidence Value function concept. That's why most of the time the system is giving the 1st answer as the correct answer. QAEdu is using the notion of an experiential knowledge base in which the knowledge has modeled from the documents provided by the teachers. Thus for the experiential knowledge base validation, we

TABLE 10.5

Unlike IR Systems, Which Uses (F, P, R), "CogQA_Edu" Uses "FHS", "FARR", "FARWR", "TRR", "RR" as a Measuring Index for the Evaluation of the Sample 30 Questions Answered by the "CogQA_Edu" System. "CogQA_Edu" Is a Cognitive Powered System so the Evaluation Process of Such Systems Should Be Like This

Q No.	FHS	FARR	FARWR	TRR	RR	Q No.	FHS	FARR	FARWR	TRR	RR
1	1	1	$\frac{1}{8}$	2.08	1	16	1	1	$\frac{1}{6}$	2.28	1
2	1	1	$\frac{1}{5}$	0.83	1	17	1	1	$\frac{1}{7}$	1.75	1
3	1	1	$\frac{1}{6}$	1.83	1	18	1	1	$\frac{1}{5}$	1.83	1
4	1	1	$\frac{1}{7}$	1.25	1	19	1	1	$\frac{1}{9}$	1.83	1
5	1	1	$\frac{1}{6}$	2.08	1	20	1	1	$\frac{1}{6}$	1.75	1
6	1	1	$\frac{1}{7}$	1.25	1	21	1	1	$\frac{1}{7}$	2.28	1
7	1	1	$\frac{1}{6}$	2.08	1	22	1	1	$\frac{1}{7}$	1.75	1
8	1	1	$\frac{1}{8}$	2.08	1	23	1	1	$\frac{1}{7}$	1.83	1
9	1	1	$\frac{1}{6}$	2.08	1	24	1	1	$\frac{1}{7}$	1.25	1
10	1	1	$\frac{1}{7}$	1.53	1	25	1	1	$\frac{1}{6}$	2.08	1
11	1	1	$\frac{1}{6}$	1	1	26	1	1	$\frac{1}{7}$	1.33	1
12	1	1	$\frac{1}{6}$	1.33	1	27	1	1	$\frac{1}{8}$	1.25	1
13	1	1	$\frac{1}{7}$	2.28	1	28	0	0	0	0	0
14	1	1	$\frac{1}{6}$	1	1	29	1	1	$\frac{1}{7}$	1.75	1
15	1	1	$\frac{1}{7}$	1.50	1	30	1	1	$\frac{1}{8}$	2.08	1

TABLE 10.6

We Have Used "RR", and "P@(k=5)" (Considered First 5 Answers) Measuring Index on the Sample "30" Questions Evaluated by "Human Experts". As "CogQA_Edu" System Implements the Multiple Answers Concept for Each "(Entity, Association)" Pair

Q No.	P@5	RR	Q No.	P@5	RR
1	0.8	1	16	0.8	0.5
2	1	1	17	1	1
3	0.8	1	18	0.8	1
4	1	1	19	1	1
5	0.6	0.5	20	0.8	1
6	1	1	21	0.4	0.25
7	1	1	22	1	1
8	0.8	1	23	0.8	1
9	1	1	25	1	1
10	1	1	25	1	1
11	0.6	0.5	26	0.6	1
12	1	1	27	1	1
13	1	1	28	0	0
14	1	1	29	0.8	1
15	0.8	0.5	30	0.8	1

have used P@k parameter, here the value of k = 5 refer Table 10.6. Because we are having multiple answers for each question, so the experts are checking the first five answers retrieved by the system, this shows an excellent result with MRR = 87:5% and Average P@5 = 84%.

10.7 CONCLUSION

Earlier Q-A systems were either proposing a new technique, talking about the upgradation of existing methods or their knowledge bases had designed in the abstraction. Thus their performance compromises as they interact in the real-world environment. This is because they are not talking about the architectural way of developing the systems which can synergize all the powers of existing techniques at one place. That's why these systems cannot perform efficiently in all the circumstances. These days everyone is talking about intelligent machines, which can perform efficiently for any condition. To achieve this particular goal we should think in the direction of a robust and scalable architecture. So to overcome the problems of earlier Q-A systems, we are proposing QAEdu a system approach based, socially inspired, cognitive agent, powered Q-A system in the field of education to teach electrical motor and basic electromagnetic concepts to the novice users. Following are the characteristics of QAEdu.

(1) QAEdu follows System approach.
(2) Currently it deals with (W=H) type questions. Shortly we will try to implement other type of questions.

(3) An NLP (Natural Language Processing) approach has implemented for the deep understanding of questions.

(4) To overcome the problems of rule-based systems, we used a dynamic approach to knowledge base construction.

(5) QAEdu is a socially inspired system (proposed) based on the system's approach. It explains the significance of experiential knowledge (provided by the teachers, who work as an expert also) for the construction of improved knowledge base.

(6) It uses the Confidence Values and Liking Values for the justification of the answer extraction process (Best & Suggested Answer).

(7) It implements the answer recommendation system (provides the suggested answer in addition to the best answer).

(8) It implements the agent's concepts for achieving autonomy.

(9) It implements the attributive searching and dictionary concepts which increase the question handling power as well as it reduces the size of the Knowledge Base.

10.8 CHALLENGES AND FUTURE ASPECTS

In this segment of the article, we are going to examine the problems that the current Q-A systems are facing and the anticipated future work to surmount these challenges. Nowadays both academia and industry are working progressively in the direction of intelligent machines, which can mimic humans. Continuous efforts were carried out since the 1950s when people were talking about Turing test (Alan Turing) to till date when we have systems like Watson (developed by IBM). Still, we are lagging somewhere. The basis behind this is the restriction and improper use of existing technology. Even though we are aware of the influence of cognitive computing approach, which synergies Speech processing, NLP, Computer vision, Machine Learning, and many more technologies simultaneously. However, we are not using its full power at full capacity. Existing systems are either recommending new techniques or talking about modifications in the existing ones, but no one is thinking of an architectural way for crafting the systems. Subsequently, to utilize the full power of cognitive computing, we require a robust and scalable architecture which can offer a comfortable and adaptive interface to synergize all these technologies within itself. At this moment, People misinterpreted the power of cognitive computing; they think that it will replace humans. The cognitive system aims to assist the humans not to replace them (cognitive computing does not talk about the meta-cognition which may be achieved in future with quantum computers). Nowadays we are using the web mining approach for answer extraction from the vast web. However, the extracted answers face the problem of accuracy and robustness. For solving answer accuracy problem, we should go with an experience-based designing of the knowledge base. In the future proposed work, we will try to provide additional features (social cognition approach to problem-solving, web mining approach for knowledge extraction and the audio-visual handling capability) to QAEdu.

10.8.1 Hardware Implementation for Providing the Embodiment to the Cognitive Agent

Figure 10.13 describes the hardware implementation for providing the embodiment to the proposed cognitive agent QAEdu. It is a cognitive agent thus it tries to mimic humans. Therefore it consists of following parts (camera, mike, speakers, servo motor, and arduino board). As the human agents consists of "eyes" for visual recognition, "ears/mouth" for speech processing, and has "neck" movement for adjusting the vision power thus by inspiring with the human agent the proposed cognitive agent's components are selected accordingly. Shortly, the "audio/visual" capacity and other capabilities can be provided to the proposed cognitive agent because of the proposed hardware.

(1) Camera mimics the "eyes" of a human agent.
(2) Mike mimics the "ears" of a human agent.
(3) Speakers mimics the "mouth" of a human agent.
(4) Servo Motor provides the movement to the "camera (eyes)" which mimics the "neck" of a human agent.
(5) Arduino-uno boards provide the controlling and interfacing to the hardware components.

FIGURE 10.13 Snapshot of the hardware designed for providing embodiment to the proposed agent "CogQA_Edu". Using this hardware interface additional capabilities can be provided to the cognitive agent.

10.9 DISCUSSION

Developing a robust and perfect system is not a single day process; it is an evolving and continuous process. Here we are trying to create a cognitive system QAEdu, which currently have the following features.

(1) **Cognitive Computing-Powered System:** Cognitive computing synergies Speech processing, NLP, Computer vision, Machine Learning, and many more technologies simultaneously under one roof. Due to all these techniques, cognitive systems try to perform the best in any situation. Cognitive systems should have the power of sense, comprehend, and act. Presently, in QAEdu, we have tried to implement all the possible techniques. Still, we have to achieve much more which we will implement shortly. Table 10.7 describes the list of cognitive features that QAEdu possesses.

(2) **Experience-based Knowledge-Base:** If we craft the knowledge base of any system in abstraction, and it has not exposed to the real-world environment. Unquestionably, such systems will not perform better in all conditions. QAEdu system is crafted deliberately under the constraints of cognitive computing approach. Thus, its knowledge base module is adaptive. It can face the real-world environment, and in the case of any doubt, it requests the teacher and updates its knowledge (as the social aspect is yet to be implemented). Therefore, we are proposing the experience-based knowledge base for QAEdu. The majority of the existing systems are using a semantic web approach for data

TABLE 10.7

Every Cognitive System Should Have the Following Three Capabilities: "Sensing", "Comprehension", and "Action". Following Comparison Table Shows the Cognitive Computing Capabilities That Are Present in the System "CogQA_Edu"

Cognitive Property	Achieved	Will be Implemented as Future Work
Sense		
Speech Processing	—	Will Implement Shortly
Audio & Visual Capability	—	Will Implement Shortly
Comprehend:		
NLP Powered	Yes	—
Evidence based Answering	Yes	—
Knowledge-Base	Yes	—
Recommendation	Yes	—
Social Cognition	Proposed	—
Act:		
System Approach	Yes	—
Agent-based System	Yes	—
Expert System	Yes	—

mining and answer extraction. The reliability of the extracted answers from such systems is very less. It always suffers from the problem of preciseness, accuracy, and authenticity. QAEdu resolves this issue by providing experiential knowledge base. (A teacher or domain expert offers the knowledge.) As we humans do, in the case of any uncertainty, we rely on domain expert rather than searching for information on the web.

(3) **NLP Power:** NLP provides the power of in-depth analysis and understanding of questions to QAEdu. We know that the clear understanding of the problem helps in extracting the exact and the best answer from the knowledge base. So here we have implemented the full power of NLP provided by SharpNLP (open source, English language natural language processing tools written in C#). It is an English language natural language processing tools written in C# and centered around a port of the Java Open-NLP library (http://opennlp.sourceforge.net/). It includes a sentence splitter, a tokenizer, a part-of-speech tagger, a chunker (used to "find non-recursive syntactic annotations such as noun phrase chunks"), a parser, a name finder, and a co reference tool.

(4) **Attributive and Dictionary-based Searching of Attributes:** In QAEdu, we have implemented dictionary concept (for synonym replacement in the questions) and the attributive searching (for linguistic components extraction from the questions) (refer to Examples 10.1 and 10.2). Since, some of the components in questions (like adjective, adverb) may change the semantics of a sentence and the synonym handling power increase the question understanding ability (for example, "Capital" and "Commercial Capital" resembles the same syntactically as both are talking about the capital of any country. However, semantically they have a different meaning. As one is talking about official capital and the other is talking about the business capital of a country).

(5) **Confidence and Liking Value-based system:** For validating answer extraction process, we are using confidence and Liking Value functions. These functions help in the selection process of the best answer and suggested answer, respectively (refer to Figures 10.3–10.6). In QAEdu, we have implemented the answer suggestion approach (like a recommendation system). We are taking feedback (as Liking Values on the scale of (−10 to 10)) from the students and store them in an XML file. Whenever a student asks the question, the system gives the best answer based on Confidence Values and provides a suggested answer by using Liking Values.

REFERENCES

[1] W. Woods (1973), *Progress in natural language understanding - an application to lunar geology*, in *Proc. of the American Federation of Information Processing Societies (AFIPS)*, 42: 441–450. AFIPS Press.

[2] Ann Copestake, Karen Sparck Jones (1990), Natural language interfaces to databases, *Knowledge Engineering Review*, 5, 225–249.

[3] Ion Androutsopoulos (1995), Natural language interfaces to databases An introduction, *Natural Language Engineering*, 1, 2981.

[4] Majdi Owda, Zuhair Bandar, Keeley Crockett (2007), *Conversation-based natural language interface to relational databases*, in *Proceedings of IEEE/WIC/ACM International Conferences on Web Intelligence and Intelligent Agent Technology Workshops*, Los Alamitos, CA, USA, November 2007, pp. 363–367.

[5] Vinitha Reddy, Kyle Neumeier, Joshua McFarlane, Jackson Cothren, Craig W. Thompson (2007), Extending a natural language interface with geospatial queries, *IEEE Internet Computing*, 11, 82–85.

[6] Anh Kim Nguyen, Huong Thanh Le (2008), *Natural language interface construction using semantic grammars*, in *Proceedings of the 10th Pacific Rim International Conference on Artificial Intelligence (PRICAI-08)*, 2008, pp. 728–739.

[7] Valentin Jijkoun, Maarten de Rijke (2006), Overview of WiQA, in *Evaluation of Multilingual and Multi-modal Information Retrieval*, Springer-Verlag, Berlin, Heidelberg, 2006, pp. 265–274.

[8] Eugene Agichtein, Chris Burges, Eric Brill (2007), *Question answering over implicitly structured Web contentIntelligence*, in *Proceedings of the IEEE/WIC/ACM International Conference on Web*, IEEE Computer Society, Washington, DC, USA, 2007, pp. 18–25.

[9] Norbert Fuhr, Jaap Kamps, Mounia Lalmas, Andrew Trotman (2008), *Focused Access to XML Documents*, in *6th International Workshop of the Initiative for the Evaluation of XML Retrieval, INEX 2007*, Selected Papers, Lecture Notes in Computer Science, Springer, vol. 4862.

[10] Pierre Zweigenbaum (2003), *Question answering in biomedicine*, in *Proceedings of the EACL2003 Workshop on NLP for Question Answering*, Budapest, p. 14.

[11] Erik Tjong, Kim Sang, Gosse Bouma, Maarten de Rijke (2005), *Developing offline strategies for answering medical questions*, in *Proceedings of the AAAI-05 Workshop on Question Answering in Restricted Domains*, Pittsburgh, PA, USA, p. 4145.

[12] D. Moll, Jos Luis Vicedo (2007), Question answering in restricted domains: An overview, *Computational Linguistics*, 33 (1), 41–61.

[13] A.P. Prajapati, D.K. Chaturvedi (2017), Ontology based knowledge representation for cognitive decision making in teaching electrical motor concepts, in Silhavy R., Senkerik R., Kominkova Oplatkova Z., Prokopova Z., Silhavy P. (eds) *Artificial Intelligence Trends in Intelligent Systems. CSOC 2017*. Advances in Intelligent Systems and Computing, 573, pp. 43–53, 2017, Springer Cham, doi: 10.1007/978-3-319-57261-15.

[14] A.P. Prajapati, D.K. Chaturvedi, (2017), *Semantic Network Based Knowledge Representation for Cognitive Decision Making in Teaching Electrical Motor Concepts*, in *2017 International Conference on Computer, Communications and Electronics (Comptelix)*, IEEE, 978-1-5090-4708-6/17.

[15] A. Bernstein, E. Kauffmann, C. Kaiser and C. Kiefer, (2006), Ginseng: A Guided Input Natural Language Search Engine, in *Proc. of the 15th workshop on Information Technologies and Systems (WITS 2005)*, pp. 45–50, MVWissenschaft, Mnster.

[16] V. Lopez, V. Uren, E. Motta and M. Pasin (2007), AquaLog: An ontology-driven question answering system for organizational semantic intranets, *Journal of Web Semantics: Science Service and Agents on the World Wide Web*, 5(2): 72–105.

[17] P. Cimiano, P. Haase and J. Heizmann, (2007), *Porting natural language interfaces between domains an experimental user study with the ORAKEL System*, in Chin, D. N., Zhou, M. X., Lau, T. S. and PuertaA. R., editors, *Proceedings of the International Conference on Intelligent User Interfaces*, pp. 180–189, Gran Canaria, Spain. ACM.

[18] C. Wang, M. Xiong, Q. Zhou and Y. Yu, (2007), *PANTO: A portable natural language interface to ontologies*, in Franconi, E., Kifer, M., May, W., editors, *Proc. of the 4th European Semantic Web Conference*, pp. 473–487, Innsbruck, Austria. Springer Verlag.

[19] O. Fernandez, R. Izquierdo, S. Ferrandez, J. L. Vicedo, (2009), Addressing Ontology-based question answering with collections of user queries, *Information Processing and Management*, 45 (2): 175–188.

[20] D. Damljanovic, M. Agatonovic, H. Cunningham, (2010), *Natural language interface to ontologies: Combining syntactic analysis and ontology-based lookup through the user interaction*, in Aroyo, L., Antoniou, G., Hyvnen, E., ten Teije, A., Stuckenschmidt, H., Cabral, L. and Tudorache, T., editors, *Proc. of the European Semantic Web Conference, Heraklion*, Greece. Springer Verlag.

[21] C. Unger and P. Cimiano, (2011), Pythia: Compositional meaning construction for ontology-based question answering on the semantic web, in Muoz, R., Montoya, A., Mtais, E., editors, *NLDB 2011. LNCS*, vol. 6716, p. 153160. Springer, Heidelberg.

[22] E. Cabrio, J. Cojan, A.P. Aprosio, B. Magnini, A. Lavelli and F. Gandon, (2012), *QAKiS: An open domain QA system based on relational patterns*, in *Proceedings of the ISWC 2012 Posters and Demonstrations Track. CEUR Workshop Proceedings*, vol. 914.

[23] V. Lopez, M. Fernndez, E. Motta, and N. Stieler, (2012), Power Aqua: supporting users in querying and exploring the semantic web, *Semantic Web*, 3(3), 249–265.

[24] C. Unger, A.C.N. Ngomo, v Gerber, and P. Cimiano, (2012), *Template-based question answering over RDF data*, in *Proceedings of the 21st International Conference on World Wide Web*, pp. 639–648. ACM.

[25] A. Kalyanpur, et al. (2012), Structured data and inference in DeepQA, *IBM Journal of Research & Development*, 56(3/4).

[26] N.N. Axel-Cyrille, L. Bhmann, C. Unger (2013), Sorry, I dont speak SPARQL Translating SPARQL Queries into Natural Language, in *International World Wide Web Conference Committee (IW3C2).WWW 2013*, May 13–17, 2013, Rio de Janeiro, Brazil. ACM, 978-1-4503-2035-1/13/05.

[27] S. Ferr (2013), *SQUALL: A controlled natural language as expressive as SPARQL 1.1*, *NLDB 2013*, LNCS 7934, pp. 114–125, 2013. Springer-Verlag, Berlin, Heidelberg.

[28] J.D. Kim, K.B. Cohen, (2013), *Natural language query processing for SPARQL generation: A prototype system for SNOMEDCT*, in *Proceedings of BioLINK SIG*.

[29] C. Pradel, O. Haemmerl and N. Hernandez, (2014), *Swip: A natural language to SPARQL interface implemented with SPARQL, ICCS 2014*, LNAI 8577, pp. 260–274, Springer International Publishing, Switzerland.

[30] H. Shizhu, Z. Yuanzhe, K. Liu and J. Zhao, (2014), *CASIA@V2: A MLN-based question answering system over linked data*, in *CLEF 2014 Working Notes Papers*.

[31] Boris Katz, Sue Felshin, Jimmy J. Lin, Gregory Marton (2004), Viewing the Web as a virtual database for question answering, in Mark T. Maybury, editor, *New Directions in Question Answering*, AAAI Press, pp. 215–226.

[32] C. Kwok, O. Etzioni, D. Weld, (2001), *Scaling question answering to the Web*, in *Proc. of the 10th International Conference on World Wide Web*, pp. 150–161, Hong Kong, China. ACM.

[33] R.D. Burke, K. J. Hammond, V. Kulyukin, (1997), *Question answering from frequently-asked question files: Experiences with the FAQ finder system*, in *Proc. of the World Wide Web Internet and Web Information Systems*, 18 (TR-97-05), pp. 57–66, Department of Computer Science, University of Chicago.

[34] O. Ferret, B. Grau, M. Huraults-Plantet, (2000), *Finding an answer based on the recognition of the issue focus*, in *Proceedings of TREC-10*, p. 39

[35] D. Laurent, P. Sgula, S. Ngre Cross, (2007), Lingual question answering using QRISTAL for CLEF 2006, *Lecture Notes in Computer Science*, 4730, pp. 339–350.

[36] Parthasarathy, S., Chen, J. (2007), *A Web-based Question Answering System for Effective e-Learning, in Abdelghani, Bouziane et al. /Procedia Computer Science 73 (2015) 366–375, Proceedings of IEEE International Conference on Advanced Learning Technologies,* 2007, pp. 142–146.

[37] B.F. Green, A.K. Wolf, C. Chomsky, K. Laughery, (1961), *Baseball: An automatic question-answering*, in *Western Joint IRE-AIEE-ACM Computer Conference*, 219224, 1961, IRE-AIEE-ACM 61 (Western), New York: ACM.

[38] A. Ittycheriah, M. Franz, W.J. Zhu, A. Ratnaparkhi, R.J. Mammone, (2010), *IBMs statistical question answering system*, in *Proceedings of the Text Retrieval Conference TREC-9.*

[39] A. Moschitti, (2010). *Answer filtering via text categorization in question answering systems*, in *Proceedings of the 15th IEEE International Conference on Tools with Artificial Intelligence*, 2003, pp. 241–248.

[40] L. Han, Z.T. Yu, Y.X. Qiu, X.Y. Meng, J.Y. Guo, S.T. Si, (2008), *Research on passage retrieval using domain knowledge in Chinese question answering system*, in *Proceedings of IEEE International Conference on Machine Learning and Cybernetics,* 5, pp. 2603–2606.

[41] K. Zhang, J. Zhao, (2010), *A Chinese question answering system with question classification and answer clustering*, in *Proceedings of IEEE International Conference on Fuzzy Systems and Knowledge Discovery (FSKD)*, 6, pp. 2692–2696.

[42] Ma Lin, Lu Zhengdong, Li Hang (2016), *Learning to Answer Questions from Image Using Convolutional Neural Network*, AAAI.

[43] Caiming Xiong, Merity Stephen, Socher Richard, (2016), *Dynamic Memory Networks for Visual and Textual Question Answering*, in *International Conference on Machine Learning.*

[44] Fukui Akira, Huk Park Dong, Yang Daylen, Rohrbach Anna, Darrell Trevor, Rohrbach Marcus, (2016), *Multimodal Compact Bilinear Pooling for Visual Question Answering and Visual Grounding*, Empirical Methods in Natural Language Processing EMNLP.

[45] Krishnamurthy Jayant, Tafjord Oyvind, Kembhavi Aniruddha (2016), *Semantic Parsing to Probabilistic Programs for Situated Question Answering*, EMNLP.

[46] Ray Arijit, Christie Gordon, Bansal Mohit, Batra Dhruv, Parikh Devi (2016), Question Relevance in VQA: Identifying Non-Visual And False-Premise Questions, EMNLP.

[47] D. Wang, H. Han, Z. Zhan, J. Xu, Q. Liu, G. Ren, (2015), A problem solving oriented intelligent tutoring system to improve students acquisition of basic computer skills. *Computers & Education*, 81, 102–112, doi: https://doi.org/10.1016/j.compedu.2014.10.003.

[48] D. Hooshyar, Ahmad, R. B., M. Yousefi, F. D. Yusop, S.-J. Horng, (2015), A flowchart-based intelligent tutoring system for improving problem solving skills of novice programmers. *Journal of Computer-Assisted Learning*, 31(4), 345–361, doi:10.1111/jcal.12099

[49] J.M. Harley, F. Bouchet, M.S. Hussain, R. Azevedo, R. Calvo, (2015), A multi-componential analysis of emotions during complex learning with an intelligent multi-agent system. *Computers in Human Behavior*, 48, 615–625. doi:10.1016/j.chb.2015.02.013

[50] B. Grawemeyer, M. Mavrikis, W. Holmes, Sergio, G. S., Wiedmann, M., Rummel, N., April 2016, *Affecting Off-taskBehaviour: How affect-aware feedback can improve student learning.* Paper presented at the *ACM international Conference Proceeding Series.* doi: 10.1145/2883851.2883936

[51] ElGhouch, N., El Mokhtar, E.N., and Seghroucheni, Y. Z. (2017), Analysing the outcome of a learning process conducted within the system ALS-CORR [LP]. *International Journal of Emerging Technologies in Learning*, 12(3), 43–56.

[52] F. Grivokostopoulou, I. Perikos, I. Hatzilygeroudis (2017), An educational system for learning search algorithms and automatically assessing student performance, *International Journal of Artificial Intelligence in Education*, 27(1), 207–240. doi: 10.1007/40593-016-0116-x

[53] M. Samarakou, P. Prentakis, D. Mitsoudis, D. Karolidis, S. Athinaios (2017), *Application of fuzzy logic for the assessment of engineering students*. Paper presented at the *IEEE Global Engineering Education Conference, EDUCON*, doi: 10.1109/EDUCON.2017.7942914

[54] B. Mostafavi, T. Barnes (2017), Evolution of an intelligent deductive logic tutor using data-driven elements. *International Journal of Artificial Intelligence in Education*, 27(1), 536, doi: 10.1007/s40593-016-0112-1

[55] Masoud I. El AghaA.M. Jarghon, S.S. Abu-Naser (February 2018), An intelligent tutoring systems for teating SQL, *International Journal of Academic Information Systems Research (IJAISR)*, 2(2), 1–7.

[56] A.C. Graesser, X. Hu, B.D. Nye et al. (2018), Electronix tutor: An intelligent tutoring system with multiple learning resources for electronics, *IJ STEM Ed* 5: 15. https://doi.org/10.1186/s40594-018-0110-y

[57] A.P. Prajapati, A. Chandiok, D.K. Chaturvedi (2019), Semantic network based cognitive, NLP powered question answering system for teaching electrical motor concepts, in Akoglu, L., Ferrara, E., Deivamani, M., Baeza-Yates, R., Yogesh, P. (eds), *Advances in Data Science. ICIIT, Communications in Computer and Information Science*, vol. 941. Springer, Singapore. doi: 10.1007/978-981-13-3582-2-8.40

[58] Cyrille Bataller and Jeanne Harris (2015, May 21), Turning cognitive computing into business value today. https://www.accenture.com.

[59] Amit Mishra, Sanjay Kumar Jain (2016), A survey on question answering systems with classification, *Journal of King Saud University Computer and Information Sciences*, 28, 345–361, https://doi.org/10.1016/j.jksuci.2014.10.007.

[60] D. Moll, J.L. Vicedo (2007), Question answering in restricted domains: An overview, *Computational Linguistics*, 33(1), 41–61.

[61] N. Indurkhya, F.J. Damereau (Eds.) 2010, *Handbook of Natural Language Processing*, 2nd ed., Chapman & Hall/CRC, Boca Raton, FL.

[62] J. Lin, B. Katz (2003), *Question answering from the web using knowledge annotation and knowledge mining techniques*, in *Proceedings of the Twelfth International Conference on Information And Knowledge Management, CIKM 03*, pp. 116–123), New York, NY, USA: ACM.

[63] F. Rinaldi, J. Dowdall, G. Schneider (2004), *Answering questions in the genomics domain*, in *Proceedings of the ACL04 Workshop on Question Answering in Restricted Domains*.

[64] Poonam Gupta, Vishal Gupta, (2012), A survey of text question answering techniques, *International Journal of Computer Applications* (09758887), 53(4).

11 Smart Cities and Industry 4.0

P.M. Kavitha
SRM Institute of Science and Technology, Chennai, India

M. Anitha
P.B. College of Engineering, Chennai, India

CONTENTS

DOI: 10.1201/9781003156123-11

11.1 INTRODUCTION

The 4th Industrial Revolution (4IR) harnesses the potential requirements for integrating the cities on the various innovation and technological concepts, making public services improvised and effective and providing 100% quality and wide infrastructure needs. Also, Industry 4.0 is more effective since the applications are more effective, which perfect and suit the world [1]. Thus, it attracts newer investment opportunities from high-tech manufacturing industries, thereby increasing the small-scale investments too. In addition, smart cities also reduce the dependencies on the overburdening infrastructure, thereby focusing on the efficiency and effectiveness of production. Sensors can be fixed to machines, which makes the evaluation of pollution tense free for the government. This can be made via Grid networks. Insights can be used to determine the logistics requirements based on the interconnection in the networks if needed. The smart city 4.0 is trying to invest in huge data connections and improve its partnerships, thus transforming and shaping future cities [3]. Figure 11.1 describes the revolution of industry [2]. Also, the massive increase in the interconnections and the wealth of the data gives way for more informed decisions and makes us produce new designs and exciting business models, which are becoming more centralized and the changes in the industrial profile, giving more activities and jobs.

Modern urbanization refers to the development of smart cities 4.0, which impacts the 4IR. Smart city development promotes smart economic background, smart population calculation, smart governance, smart living, intelligence infrastructure, etc. The model and the development in the newest urbanization follow the 4IR, and its wholesome applications to industry 4.0 [4]. Various industries and processes

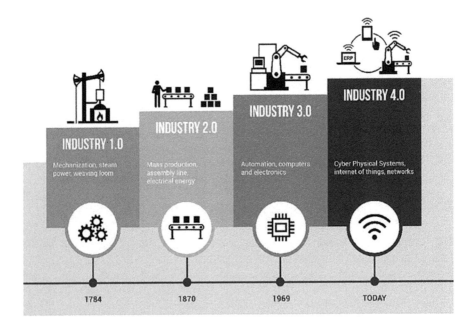

FIGURE 11.1 Industrial Revolution [5].

which follow IR 4.0 include the Internet of Things, cyber-physical systems, cloud computing, and big data. Several other industries include the plans of construction, managing, and sustainable development in cities. The most challenging task among these includes integrating and interaction between the new cities to be developed and the new urbanization. The technologies for industrialization are to create a brand new structure for the greater city and the future world. With these new technologies, newer problems may be solved in resource utilization, energy efficiency management, environmental and geographical changes, and cities' production [6]. The city's developmental process of city 4.0 and smart infrastructure make way for new industrialization, a new way of living, various employment opportunities, and education. It also paves the way for social and human capital and its attraction towards the financial resources for new business opportunities.

11.1.1 INTRODUCTION TO SMART CITY 4.0

The new role for leadership in global social, economic, and ecological sustainability are explored with the smart city development. This considers the population, consumption, technology, diversity, non-violent changes in the economy too. Also, the development includes the policies, programs, and plans that eventually address the policies, programs, and plans associated with the issues, special attention to the international disaster mitigation and recovery planning, sustainable development, and regenerative design [7, 8]. A smart city will work with multiple technological solutions to manage the city's assets securely. The city's assets include the local departments, including health care, transportation and library management, education departments, law and enforcement and power generation [7], and other such services for the community [9]. A smart city doesn't only mean improving the infrastructure but also the quality of living with various technological services and citizens' needs.

Technology and innovation are the two factors essential for the smart city process. Figure 11.2 represents the components of smart cities. Technology is driven by

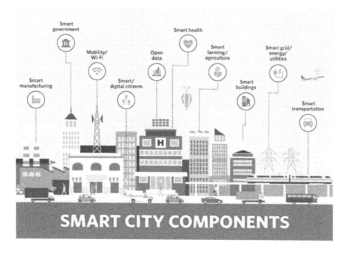

FIGURE 11.2   Components of Smart City [10].

business, and innovation is driven by large-scale urbanization. Technology takes its way by city officials to interact among themselves and the new community to develop its infrastructure. Data is collected from various sources, including the citizens and the objects. This is also achieved with the use of sensors and real-time systems. Inefficiency in this process will be tackled with the information and knowledge gathered from the sources of the city being evolved, and it derives a better place for the citizens to live in. ICT, the Information and Communication Technology, makes sure to improve productivity and make cities more interactable [11]. This reduces the costs and the utilization of resources in the right way. This, in turn, improves the contact between citizens and the government. The transportation field, traffic management, health sector, energy sources, water management, and new modern agricultural and waste product management improve and develop as a smart city. Smart city applications are developing in hands with urban flows that allow real-time responses and challenges. A transnational relationship with the citizens is established, and it paves the way to respond to the challenges [12].

11.1.2 Smart City vs. Industry

The transformation of a society that will be greatly urbanized brings huge irreplaceable problems, which is the most challenging task in smart city development. Modern citizens who seem to be the host for the problems have paved the way for mental framework solutions [13]. Modern technologies connect and improve the lives of the residents and the upcoming businesses. This entire framework defines the smarter city 4.0, where modernized technologies and factors drive a newer and efficient lifeline for the cities around the world. A huge inspiration lies in creating a smart city, where the key components lie on the infrastructure and environmental, emergency, response, traffic and mobility management, and few others [14]. These technologies get integrated and their applications are put together to extract the best from them. The improvement in the efficiency of emergency cases and their responses is the key to IoT's successful smart city. IoT stands to be the devices connected physically, including the robot, vehicles, and infrastructural things connected to exchange the data [15]. Thus, it creates new opportunities and growth on the physical and digital media to improve efficiency and livelihood. The biggest challenge lies in finishing the project, as slowing down creates overuse of the resources and human effort. Artificial Intelligence (AI) is the advanced technology that strives to enable the 4IR with its advanced network coverage [16]. The 5G coverage, along with the edge and cloud technology, gives support to millions of smart city with the sensing devices.

The software platform connects the bridge between the technologies like 5G, Edge computing, AI, and ML [17]. It offers all the necessities for igniting smart city development, which includes the areas like Supply Chain Management and Logistics & Telecommunications industry. Under the supply chain management technology, the citizens' lives rely on the smart city requirements where the industrial revolution 4.0 takes a huge step. This will make the future world cost-effective in many sectors and connected. The Logistics industry is trying to improvise the USD 43.9 billion in 2017 to USD 83 billion in 2023, where the CAGR is around 12.9% [18, 19]. This, on the other hand, offers several growth opportunities to the citizens. The Telecom area is where a high impact is seen [15]. The connectivity cases are the prominent features to remove and improvise the risks in industrialization. Smart cities play an important role in the development of smart city coverage. IoT is a vast area, which helps to increase the number of public and private organizations. These organizations will wisely invest in smart city development through cyber-physical systems, which results in the 4IR [20].

11.2   4IR

The ultimate idea behind having partnerships with big bodies makes the cities transform into smart cities. The 4IR transforms the cities technologically and also the lives of millions of people around there. Henceforth, the new technologies improve the existing infrastructure's function, which significantly alters the future methodologies of urbanization [21–24]. The 4IR is completely responsible for energy efficiency and reliability. A sustainable plan must be created to incorporate a smart city. The 4IR makes a huge change in the fundamentals of the cities by transforming (Figure 11.3) [25].

The citizen's life in living, the way of work are related to each other. In the 4IR, a new chapter of human development will be established with extraordinary technological advances, which combines the factors from the 1st, 2nd, and 3rd revolutions. These advances combined with the physical, digital, and biological worlds make a promising factor. The speed, depth, and depth of the revolution make countries and cities worldwide think about the development and how the organizations are created with values and importance. Technology-driven changes are more seen in this 4IR [26]. It provides an opportunity for the residents, including leaders, policy-makers, and people from all backgrounds irrespective of the place and way. Thus, the 4IR seems to become human-centered and human-specific in the future. More importantly, beyond technology being developed, the ability for the citizens to live positively is to be produced [27–29].

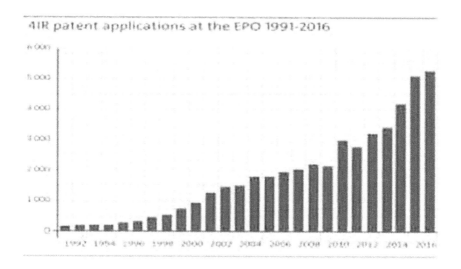

FIGURE 11.3   4IR Application [30].

11.2.1 Four Concepts on Industry 4.0

Each development and industrialization will have its own set of factors to prove a step or portfolio on which the city will be built. Other than the technological advances that involve in smart city development, the key to the revolution are listed below (Figure 11.4) [31].

11.2.1.1 Educating People

We know machines don't take over jobs, but routine tasks can accomplish them. But, to achieve this, a higher level of qualification is required, which must be focused on. Thus the 4IR gives development and focuses on educating minds [20, 32]. By this, a new and great city can be developed. This makes a chance for employment and enhances the professional career, which is the most important factor above all, as it shapes the minds of the young. Anything and everything can be achieved with the help of ideas and innovations where education is essential in building them [33].

11.2.1.2 Organizational Transformation

Every sector in an organization will have several departments to organize the work in the project [34]. Each department must strengthen itself to produce a big and better outcome. Also, the cross-sections in the departments make it more easy and convenient for the officials to extract the best from them [35, 36]. Along with the existing department, several new technologies will be crafted according to the new requirements, creating new employment opportunities [37–40]. The various departments in any organization must be cross-sections, making the data and other such processes more relaxing. This also allows for lesser hierarchical environments.

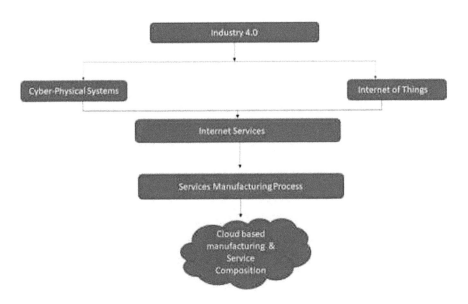

FIGURE 11.4 The framework of industry 4.0 [41].

11.2.1.3 Decentralization of Work Networks

Machinery does most of the work in this era. Thus, having some good machines and robots makes things easier, productive, and less time consuming. Having just simple machinery which is effective will do the process [42]. Investing in the components will be focused, which saves a lot of pence. The manufacturing sector, by the way, is developed, and new resources will be essential in the process. Thus, new methodologies will be used to reduce the time factor and the manufacturing costs [43].

11.2.1.4 Sustainability

A better city must be a place for better lives. Pollution will be the major cause that affects human lives and creates problems [44]. The technological revolution tries to reduce pollution to the maximum with tremendous production. Having an eye on every process conserves the raw materials. It also provides a sustainable and efficient society. Today's conservation of raw materials remains for tomorrow's world and the future. These improvisations give an add-on benefit to the citizens and provide a better way of living. Also, the industries with developed factors aim to bring out the maximum with limited utilization of resources [45–47].

11.2.2 INTEGRATION OF 4IR TECHNOLOGY

Thanks to the global revolution, technology is connected and fits in everyone's hand. Various new technologies combined with innovations and concepts make the world effective. This provides a betterment in the lifestyle of the citizens [48]. Also, it creates more employment which is a major challenge today.

In the beginning, the industrial revolution made a new energy source, i.e., steam, which became the major source of energy. This allowed the industries to move forward on mechanization and improving the various other technologies to improve productivity. In the 2nd industrial revolution, electricity, the major innovation of the era, was made [49]. In addition to that, various other innovations were made in this era: new business opportunities and various other industries. This created employment for many citizens. The 3rd industrial revolution began in the late twentieth century [50]. This era brought a major change in information technology. The most important era is this era, which also focused on the production processes. The biggest of the industrial revolution in history is the 4IR. This is the milestone for all the other revolutions as it connects the technologies into societies and industries.

The 4th revolution created the connectivity of all the emerging technologies. Thus, the digitalization of the industries is speeded up [51]. The cloud platform and other such 5G technologies were technologies that created a boom in the industries. Robotics, Artificial Intelligence (AI), and Internet of Things (IoT) are the few developed technologies that helped the innovation occur [52, 53]. This 4th revolution not only invented technology but will also boom in many fields. This enhanced connectness and productivity on the major social issues. A new and versatile level for the businesses with innovative and agile models will be created. The innovation and integration were constantly built-in on a big scale and in small businesses too (Figure 11.5).

The Fourth Industrial Revolution Is in Full Swing

Percentage of Customers Who Believe the Following Technologies
Will Transform Their Expectations of Companies

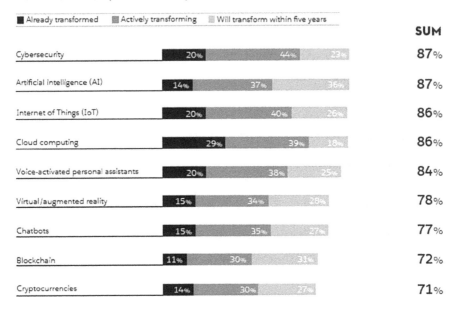

	Already transformed	Actively transforming	Will transform within five years	SUM
Cybersecurity	20%	44%	23%	**87%**
Artificial intelligence (AI)	14%	37%	36%	**87%**
Internet of Things (IoT)	20%	40%	26%	**86%**
Cloud computing	29%	39%	18%	**86%**
Voice-activated personal assistants	20%	38%	25%	**84%**
Virtual/augmented reality	15%	34%	28%	**78%**
Chatbots	15%	35%	27%	**77%**
Blockchain	11%	30%	31%	**72%**
Cryptocurrencies	14%	30%	27%	**71%**

FIGURE 11.5   Revolution swing of the industry [54].

11.2.2.1 Platform Capabilities

The platform that incorporates No-code and Low-code develops the businesses with the help of these software applications with the help of codes. This is the most convenient way to develop and use the software. Built-in functional components available in the library are created with the help of visual modeling methodology [55]. The components can just be selected and can be added to the application as per the workflow. Thus, greater efficiency is achieved by the developers in a faster way. This makes great progress for all the people of all levels in the business, saving a huge sum of money used in the maintenance. Also, the platforms are made available in all packages, which the developers can choose [56–58].

11.2.2.2 No-Code Development Capabilities Scenario

The No-code development scenario provides users who solve their basic crisis and will offer a standard and straightforward solution. SMBs take the credit, as these will be more suitable to integrate the challenges without a professional developer as the applications are built without any feature application usage [59].

11.2.2.3 Low-Code Development Capabilities Scenario

The Low-code capabilities scenario caters to the technology governance requirements of the enterprises. A more formalized method with scalable and approachable architectural structures that are flexible in the implementation is developed. Platform extension is also possible by using an open API. Integration with complex means and with comprehensive structures present in the libraries gives a way to incorporate with the newer and reliable technology for the future world via open source. These platforms are created to give more sophisticated applications that run on several different departments and areas. The developers need the best control on these as quality testing and analyzing are vital. This is achieved through various tools that check individual performance and detect high productivity and high speed in all possible cases.

11.2.2.4 API-Based Ecosystem

The platforms were created to integrate technology with the power of APIs. This made it easy for the developers to adapt and create new technologies hassle-free easily. With bigger business, the requirements also grow bigger and wider [60]. The enterprises with the applications are set to work across several departments and domains. Thus productivity is improved according to the next-generation technologies. The challenges faced by the modernization of legacy systems are developed in adaptation with the cloud platform. The APIs prove to be an intermediate part that stands between the cloud platform and the on-prem platform. Thus, enabling all the other applications to maintain a neutral platform installed to maintain access to the data and other services. The APIs and integration platforms work together efficiently and manage the products and their data, enabling greater deployment and efficiency. A lot of time is saved along with the cost-reduction factors. More importantly, the errors generated from the complex coding, which is the most tedious task, are made easy.

11.3 THE RIGHT DATA PLATFORM

It is essential to choose the right platform for building a great city with all the modernization factors. The right platform gives a renewable city adapted by all the residents in a hassle-free way. One such fastest growing technology is machine learning, a subset of AI, which is user-friendly and suits today's world. It has very fast growth and adapts according to the future world. Machine learning is the subset of AI, processes large, complex, and multidimensional data using applied computational techniques [61]. ML helps in building data models by predictions from the existing patterns on the data set. This enhances the decision-making strategy.

11.3.1 Machine Learning into 4IR

ML uses its algorithms in the implementation and enhances the future strands. The algorithms are classified as supervised, unsupervised, and reinforcement learning. The supervised learning algorithms are used in training the data set, which contains dependent and independent variables and predicts the target labels. Also, the prediction is based on the target data with minimal deviations from the original or the actual data. These algorithms are used in various real-time applications like quality monitoring and prediction. The input variables for these will be temperature, pressure, density, and so on. And the target will be of quality or quantity of the product. Also, supervised learning varies in the health and care sector by predictive maintenance (PdM) [62]. Depending on the algorithm used, either regression or classification, various metrics will be used to measure and maintain the model's accuracy. Thus, the ML model's performance and efficiency are improved in cross-validating the data or subsequent training.

In unsupervised learning algorithms, the basic structure of the data is analyzed first, which includes identifying the obtrusive, instinctive, and erratic or masked patterns in the input sets. These algorithms are used in the manufacturing industry, where the clustering algorithm is used in the cluster production process [63]. It determines the hidden patterns in the supply chain and logistics. It reduces the end-user's risk and is compatible with the market characteristics such as pricing, quantity, etc.

The ML has many practical applications in the manufacturing industry is exclaimed using the data on the MES, which is the Manufacturing Execution System or the SCADA - The Supervisory Control and Data Acquisition systems. ML is the most effective platform used in smart city development. Various other technologies are also used in the development as the innovation and improvements for a city cannot be afforded with a single technology.

11.3.2 BIG DATA AND IoT

For green and safe city, the water, air, and great mobility, and effective public services must reach the very last citizen. Thus, with bid data and IoT, it is a lot easier. Sensors, applications for the cities to be effective like smart water meters, which measure the accurate water quantity to every household or industry, and alters for water leakage and overflowing, water contamination or pollution can be examined easily with these technologies (Figure 11.6).

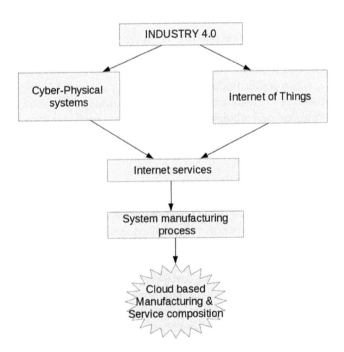

FIGURE 11.6 Impact of Industry 4.0 in various domains [64].

A very large and live data set is essential in calculating the details for a city's advancement. It includes the citizen's count, water bodies, types, and quantities of houses and factories, etc. IoT plays a very vital role in the calculation. This also helps in sorting, analyzing, and processing the data which is collected from IoT. The examples of IoT include sensor, transportation, lightings, and waste management [65].

In certain cities, smart street lights are installed around the world. These lights contain Wi-Fi hotspots with intelligent features along with a surveillance camera and a charging point where phones and electric cars can be charged. The quality of the air can also be checked, which in turn improves the air condition. The major factor to be concentrated must be on the managing and recycling of the wastes. It can consist of human waste, electrical waste, and micro wastes. Proper waste management must be implemented in the cities for a cleaner and better world. Reducing noise pollution is also important in addition with the waste management. The trash is collected as degradable and non-degradable waste for decomposing and recycling into newer products. The trashes are sent into large pipes with high pressure to the recycling units. The transportation which connects all the major cities is highly essential. Traffic management must be highly concentrated to avoid any disturbances. Thus sensors can be used to predict the traffic among cities to avoid delays. Sensors play a main role in the development of smart cities as it is used in every system. Sensors are used in traffic lights, cars, houses, and industries and every device [66, 67]. This helps in interconnecting the city with other cities. Power is the major factor, and every developed city has its way to use the power efficiently. Smart grid technologies are implemented in managing high power consumption. Data analytics can be used in the prediction of power usage effectively and distribution is managed effectively. Solar plants and panels can be used as an alternative source of energy.

The infrastructure for a smart city requires a much clear and efficient traffic and transportation method. Big data and IoT are used in this process to provide the residents with faster and smooth transportation without the need to wait for signals and other disturbances such as roadblocks and repairs. This is essential as the traffic flow in a public transportation system must be organized. The optimization of traffic is ideal and monitoring the systems needs to be done. Though this seems to be similar to the traffic management system, the control of the vehicles must be noted by the lanes and cross paths to avoid accidents [68, 69]. A real-time display at the bus stops and train stations is put, which gives the citizens an idea of the arrival and departure time of the vehicle. Also, the routes and the links, if any, are also displayed for ease. The most important and effective means is to introduce a fare card, with which the citizens need not weight at the ticket collecting centers as this saves a lot of time.

11.4 BENEFITS OF SMART CITY 4.0

Developing a smart city 4.0 includes various technological transformations. This hugely transforms the cities and the impact remains to be optimistic. The citizens will have a better way of living as it gives more employment opportunities and the education sector grows widely [70]. The physical, structural, and mental changes of the cities and the residents remain good, providing a better and greater future.

Furthermore, there are numerous advantages of building a smart city that improvises the occupants' lives from various perspectives. Out of a couple of prominent and extraordinary benefits of a smart city are listed.

11.4.1   DATA-BASED AND TRANSPORTATION

Having built a smart city, it is hassle-free for the officials to collect and revamp the data and analyze it for further developments. Also, a well-designed data analytic strategy leads to monitor the city from time to time and increases the service levels rapidly [71, 72]. With smart cities, the world shrinks and is connected with every other city, which drastically improves the efficiency of the city. It develops traffic management and helps in tracking the moving location of buses and trains. This helps the citizens in the growing population.

11.4.2   SAFER COMMUNITIES IN THE FUTURE

A smart city must for sure be a safer city too. Thus, the industrialization aims to focus on the reduction of criminality in the city with various innovative features, thereby helping the Police Department to find the offenders and sinners at a quick rate. This also includes the applications like recognition of license plates, shooting identifications, and crime centers hassle-free [65]. Natural resources are to be utilized more efficiently, thanks to scarcity. Smart technologies help cities with all the necessary steps and factors to effectively use and conserve resources, especially water and electricity.

11.4.3   REDUCTION OF THE ENVIRONMENTAL FOOTPRINT AND INCREASE IN DIGITAL EQUITY

New tools for buildings with high energy efficiency, sensors to test the air quality, and proper utilization of energy utilization are imported to improve the ecological impact positively [73]. High-speed Internet must be ensured, which proposes digital equity to the citizens. Public Wi-Fi access is a boon to the citizen, and the city must offer high-quality Internet services for all its citizens.

11.4.4   NEWER AND GREATER ECONOMIC OPPORTUNITIES

A vast improvement in regional and global competitiveness will be made with the smart city. This way, it attracts new residents and creates new business opportunities [74]. Having an open data platform in hand, which is used to access the city's information, the organizations provide more informed decisions with the help of data analysis via integrated technologies.

11.4.5   INFRASTRUCTURE IMPROVEMENT

The demonization of an older building, estates, roads will always need high investment, and the maintenance, in turn, grows largely [75]. Thus, with the help of industrialization, the cities are built with zero or minimal maintenance as they are built with predictive analytical measures. This gives a warning or notification on any repair or wear and tear of the built models.

11.5   LIMITATIONS OF SMART CITY 4.0

Every innovation and development has a boon and a bane. The boon of a smart city is already being discussed. The bane and few disadvantages are now to be seen. With the disadvantages, the researchers can develop and remodel the city. With many technologies being used in smart city creation, the tension will be developed as the usage of technologies in information and communication is extensive. Also, a question of the green city will arise, as with more innovations, the space of the green world is somehow forgotten, which is to be focused by the scientist and environmentalist to protect the world. The smart city will surely have lots of chips, sensors, cable wires, optical fibers, and many other physical things, making the environment weak. These are mostly non-degradable and produce an impact on society. Thus degradable products must be focused on, which can be recycled to get newer products. The electricity consumption for all the major needs will be vast as the city needs to be connected. Thus, the utilization of solar panels and the concept of green buildings and other renewable sources must be taken into practice. Other environmentally friendly ideas must be put forward and the innovations must be appreciated and the resources cannot become extinct. After energy efficiency, the cost is the major input that is the most required in building a smart city. The cost must be reduced in building the components and the electronic wastes must be reused to reduce the financial needs for the cities.

11.6   CONCLUSION AND FUTURE ASPECTS

Developing the smart city includes investments for new technologies and setting up the city and lies in implementing and the solution that covers real-life problems. The focus is on improving the city operations and focusing on the quality and standard of living. The technology and the devices must be a helping hand and should deal with all the major problems in the city. The smart city 4.0 is built to give a great living to the citizens and improve the same quality. A smart city, in all, must support sustainable development since it proves to be the basic aspect and nothing can be compromised in it. It deals with all the natural resources and must be cautious as they may lead to a depleted crisis. Many aspects, such as citizen-centric, pluralistic, holistic, must focus on improving the services and solving urban issues. This aims to improve the services and the solutions to the urban problems in a more effective and cost-efficient way. In the longer run, the technology must be improved and solve the city's physical services and infrastructure, thereby making every city a smart city. The 2011 census reports that nearly a population of 31% is still in urban areas and it is estimated that by 2030 almost 50% must move on to urban as a result of industrialization. This urbanization process is slow and takes time to bring the whole world into urban cities. Thus, on the other hand, a vast need for electricity, water, and transportation must be provided for the citizens, which must be concentrated. Also, the cities must be safe places to stay in. As a result, a green city with minimal waste disposal and a clean and safe city will be free to use by the citizens. With IoT, the cities are connected with major infrastructure and services. The quality of life is improved with smart traffic, smart waste disposal, health facilities, and so on. Smart cities provide sustainable growth which in turn reduces the energy cost by implementing natural resources. This development takes time to attain its full potential. But very soon, the future world will eventually witness the changes.

REFERENCES

[1] A. Safiullin, L. Krasnyuk and Z. Kapelyuk, *"Integration of Industry 4.0 technologies for "smart cities" development"*, IOP Conference Series: Materials Science and Engineering, Volume 497, *International Scientific Conference "Digital Transformation on Manufacturing, Infrastructure and Service"*, November 21–22, 2018, Saint-Petersburg, Russian Federation, Russia, IOP Publishing Ltd.

[2] www.researchgate.in

[3] R. E. Lucas, Jr, *Lectures on Economic Growth*, Cambridge: Harvard University Press, 2002, pp. 109–110, ISBN 978-0-674-01601-9.

[4] https://mobility.here.com/learn/smart-city-mobility/smart-city-technologies-role-and-applications-big-data-and-iot

[5] www.salesforce.com

[6] https://unmanned.life/2019/03/unmanned-life-the-link-between-industry-4-0-and-smart-cities/, March 2019

[7] M. Hermann, T. Pentek and B. Otto, "Design principles for industrie 4.0 scenarios", February 3, 2015.

[8] J. Schlick, P. Stephan, M. Loskyll, and D. Lappe, "Industrie 4.0 in der praktischen Anwendung", in Bauernhansl, T., M. ten Hompel and B. Vogel-Heuser, eds., *2014: Industrie 4. 0 in Produktion, Automatisierung und Logistik. Anwendung, Technologien und Migration*, 2014, pp. 57–84.

[9] Image Source, https://techstartups.com/wp-content/uploads/2020/01/Industry-4.0.jpg

[10] https://www.epo.org/news-events/in-focus/ict/fourth-industrial-revolution.html

[11] H. W. Lin, S. V. Nagalingam, S. S. Kuik and T. Murata, "Design of a global decision support system for a manufacturing SME: Towards participating in collaborative manufacturing", *Int. J. Prod. Econ.*, vol. 136, no. 1, pp. 1–12, Mar. 2012.

[12] SAP AG, "How to prepare for the fourth industrial revolution", Online: http://global.sap.com/community/ebook/2013_11_28302/enUS/index.html#/page/, Oct. 1, 2013.

[13] B. Evjemo, S. Akselsen, D. Slettemeas, A. Munch-Ellingsen, A. Andersen and R. Karlsen. ""I expect smart services!" User feedback on NFC based services addressingeveryday routines", [online]. s. 118 [cit. 2015-12-07]. doi: 10.1007/978-3-319-19743-2_18.

[14] A.A. Goncharenko, L. L. Voronova, V. I. Voronov, A. A. Ezhov and D. V. Goryachev, *"Automated support system designing for people with limited communication"*, *2018 Systems of Signals Generating and Processing in the Field of on Board Communications*, p. I-7, 2018.

[15] A.S. Trunov, L. L. Voronova, V.L. Voronov and D.P. Ayrapetov, *"Container cluster model development for legacy applications integration in scientific software system"*, *IEEE-International Conference "2018 Quality Management Transport and Information Security Information Technologies" (IT & QM & IS-2018*, pp. 815–819.

[16] C. Alippi: *Intelligence for Embedded Systems*, New York: Springer Verlag, 2014, 283pp, ISBN 978-3-319-05278-6.

[17] J. Lee, B. Bagheri and H.-A. Kao, "*Recent Advances and Trends of Cyber-Physical Systems and Big Data Analytics in Industrial Informatics*", in *IEEE Int. Conference on Industrial Informatics (INDIN) 2014*, Brazil, 2014.

[18] V. Koch, S. Kuge, R. Geissbauer, and S. Schrauf, "Industry 4.0: Opportunities and challenges of the industrial internet", PwC and Strategy, 2014.

[19] L. Heuser, Z. Nochta, N.C. Trunk, *ICT Shaping the World: A Scientific View*, ETSI, London: Wiley Publication, 2008.

[20] R. Rajkumar, I. Lee, L. Sha, J. Stankovic, *Proceedings of the 47th Design Automation Conference (DAC '10)*, ACM, New York, pp. 731–736.

[21] O. Doleski, "The integrated business model: An applied approach", in *Integrated Business Model. Essentials*, Wiesbaden: Springer Gabler, 2015, [online] Available: https://0-doi- org.ujlink.uj.ac.za/10.1007/978-3-658-09698-4_3.

[22] K. Thompson and R. Kadiyala, "Leveraging big data to improve water system operations". *16th Conference on Water Distribution System Analysis, WDSA 2014. Proc. Eng.*, vol. 89, pp. 467–472, 2014.

[23] H. Chan, M. J. Cherukara, B. Narayanan, T. D. Loeffler, C. Benmore, S. K. Gray, et al., "Machine learning coarse grained models for water", *Nat. Commun.*, vol. 10, pp. 329, 2019.

[24] S. G. Tzafestas, "Modern manufacturing systems: An information technology perspective", in *Computer-Assisted Management and Control of Manufacturing Systems. Advanced Manufacturing*, London: Springer, 1997.

[25] https://www.weforum.org/projects/fourth-industrial-revolution-andenvironment-the-stanford-dialogues

[26] A. Reiner, "*Industrie 4.0 - Advanced engineering of smart products and smart production*", *19th International Seminar on High Technology*, Piracicaba, Brasil, October 9, 2014.

[27] C.A. Valdez, P. Brauner, A.K. Schaar, A. Holzinger, M. Ziefle, "*Reducing complexity with simplicity-usability methods for industry 4.0*", *19th Triennial Congr. Int. Ergon. Assoc.*, IEA, 2015.

[28] L. Bartevyan, "Industry 4.0 – Summary report", L., DLG-Expert report 5, 2015.

[29] J. Nasser, "*Cyber physical systems in the context of Industry 4.0*", in *Automation, Quality and Testing, Robotics, 2014 IEEE International Conference on*, IEEE, Romania, 2014.

[30] https://www.weforum.org/focus/fourth-industrial-revolution

[31] https://www.iot-now.com/2019/09/06/98516-exploring-benefits-building-smart-city-country-part-two/, July 9 2019

[32] A. Cardenas, S. Amin, S. Sastry, *The 28th International Conference on Distributed Computing Systems Workshops*, Beijing, 2008, pp. 495–500.

[33] L.l. Voronova, R.V. Tolmachev and V.A. Usachev, "*Resource development to prevent riots at mass events*", *2018 Systems of Signals Generating and Processing in the Field of on Board Communications*, pp. 1–5, 2018.

[34] C. Baur and D. Wee, "Manufacturing's next act", 2015, Available at: http://www.mckinsey.com/business-functions/operations/ourinsights/manufacturings-next-act, March 10, 2017.

[35] S. Jeschke, C. Brecher, H. Song and D. B. Rawat (Eds.), *Industrial Internet of Things and Cyber Manufacturing Systems*, Springer International Publishing, 2017, pp. 3–19.

[36] A. Gilchrist, *Industry 4.0 – The Industrial Internet of Things*, New York: Springer, 2016.

[37] V. Koch, S. Kuge, R. Geissbauer, S. Schrauf, "Industry 4.0: Opportunities and challenges of the industrial internet", 2014, Available at: http://www.strategyand.pwc.com/reports/industry-4-0, March 10, 2017.

[38] H. W. Lin, S. V. Nagalingam, S. S. Kuik, T. Murata, *Int. J. Prod. Econ.*, vol. 136, no. 1, 1–12, 2012.

[39] A. Cardno, "6 critical ideas behind the smart factory and internet of things", Available at: https://blog.vizexplorer.com/6-critical-ideas-behindthe-smart-factory-and-internet-of-things-iot/.

[40] J. P. Müller, W. Ketter, G. Kaminka, G. Wargner, N. Bulling (Eds.), *A Multiagent System Perspective on Industry 4.0 Supply Networks*, Multiagent Systems Technologies, 2015.

[41] https://www.mydbsync.com/blogs/how-integration-platform-is-driving-the-4th-industrial-revolution/

[42] D. Uckelmann, "Definition approach to smart logistics", Wireless Advanced Networking, Russia, September 5, 2008.

[43] KPMG, "The factory of the future: Industry 4.0 – the challenges of tomorrow", April 17, 2016.

[44] McKinsey Digital, "Industry 4.0 How to navigate digitization of the manufacturing sector", 1–62, April 1, 2015.

[45] DHL Trend Research, "Internet of things in logistics. A collaborative report by DHL and Cisco on implications and use cases for the logistics industry", 2015.

[46] S. Schrauf and P. Berttram, "Industry 4.0: How digitization makes the supply chain more efficient, agile, and customer-focused". PWC Report, 2016. Available at: http://www.strategyand.pwc.com/media/file/Industry4.0.pdf, March 3, 2017.

[47] L. Atzori, A. Iera, G. Morabito, *Comput. Netw.* 54 (15) (2010) 2787–2805.

[48] E. Qin, Y. Long, C. Zhang, L. Huang, "Human interface and the management of information. Information and interaction for health, safety, mobility and complex environments", in *International Conference on Human Interface and the Management of Information-HIMI 2013*. pp. 173–180, 2013.

[49] S. Muntone, *Second Industrial Revolution*, The McGraw-Hill Companies, Education.com. October 14, 2013.

[50] J. Rifkin, *The Third Industrial Revolution: How Lateral Power Is Transforming Energy, the Economy and the World*, 1st ed., New York: Palgrave Macmillan, 2011, 291 s. ISBN 9780230115217.

[51] B. Guo, Z. Yu, D. Zhang and X. Zhou, "Cross-community sensing and mining", *IEEE Commun. Mag.*, vol. 52, no. 8, pp. 144–152, Aug. 2014.

[52] D.N. Bezumnov and L.I. Voronova, "Evaluation of time characteristics of real-time tasks on Arduino Uno", *Telekommunikatsii I informatsionnyye tekhnologii [Telecommunications and information technology] Moscow MTUCI*, vol. 4, no. 2, pp. 51–54, 2017.

[53] "The training stand Technology automation", [online] Available: http://smartgorod.com/rusloborudovanie/4-uchebnyi-stend-imitatsiyal.

[54] www.cadm.com/the-fourth-industrial-revolution. May 18, 2018

[55] Y. S. Yilmaz, M. F. Bulut, C. G. Akcora, M. A. Bayir and M. Demirbas, "Trend sensing via Twitter", *Int. J. Ad Hoc Ubiquitous Comput.*, Switzerland, vol. 14, no. 1, pp. 16–26, 2013.

[56] T. Ludwig, C. Reuter, T. Siebigteroth and V. Pipek, "*Crowd Monitor: Mobile crowd sensing for assessing physical and digital activities of citizens during emergencies*", *Proc. 33rd Annu. ACM Conf. Human Factors Comput. Syst.*, pp. 4083–4092, 2015.

[57] S. Ji and T. Chen, "Incentive mechanisms for discretized mobile crowdsensings", *IEEE Trans. Wireless Commun.*, vol. 15, no. 1, pp. 146–161, 2016.

[58] T. Soyata, H. Ba, W. Heinzelman, M. Kwon and J. Shi, "Accelerating mobile cloud computing: A survey", in *Communication Infrastructures for Cloud Computing*, Hershey, PA, USA: IGI Global, pp. 175–197, Sep. 2013.

[59] https://www.arcweb.com/technology-evaluation-and-selection/industrial-internet-things-iot-edge-software-platforms

[60] J. Sorensen, "Review of existing literature and deployment tracking surveys decision factors influencing ITS adoption", U.S. Department of Transportation, April 2012.

[61] M. Pouryazdan and B. Kantarci, "The smart citizen factor in trustworthy smart city crowdsensing', *IT Pro.*, 2016, [online] Available: http://nextconlab.academy/PouryazdanKantarci-IT-Professional2016

[62] F.-J. Wu and H. B. Lim, "Urban mobility sense: A user-centric participatory sensing system for transportation activity surveys", *IEEE Sensors J.*, vol. 14, pp. 4165–4174, Dec. 2014.

[63] "NIST. SP 800-14: Generally accepted principles and practices for securing information technology systems", 1996 [online]. NIST. Available at: http://csrc.nist.gov/publications/nistpubs/800-14/800-14.pd.

[64] https://blogs.uoregon.edu/smartcities/, 2003.

[65] J. Macaulay, L. Buckalew, G. Chung, *Internet of Things in Logistics*, Germany: DHL Customer Solutions & Innovation, 2015.

[66] M. Goodrich and R. Tamassia, *Introduction to Computer Security*, 2nd ed., Addison-Wesley Professional, 2014.

[67] S. Bosworth, M. E. Kabay, *Computer Security Handbook*, 4th ed., Wiley, 2002.

[68] ENISA, "The new user's guide: How to raise information security awareness", Luxembourg, 2010.

[69] M. Bishop, *Introduction to Computer Security*, 2nd ed., Addison-Wesley Professional, 2004.

[70] ISO/IEC_JTC1. ISO/IEC FDIS 27000, "Information technology - security techniques - information security management systems - overview and vocabulary", ISO Copyright Office, Geneva, Switzerland, 2009.

[71] S. Hasan, N. Siddique and S. Chakraborty, *Intelligent Transport Systems. 802.11-Based Roadside-to-Vehicle Communications*, New York: Springer Publication, 2013.

[72] H. Hartenstein and K. Laberteaux, *VANET Vehicular Applications and Inter-Networking Technologies*, 1st ed., London: Wiley Publication, 2010.

[73] W. Sarni, C. Stinson, A. Mung, B. Garcia, S. Bryan and J. Swanborough, "Harnessing the fourth industrial revolution for water. fourth industrial revolution for the earth series", *World Economic Forum*, 2018.

[74] H. Carvalho, V. Cruz-Machado, *Integrating Lean, Agile, Resilience and Green Paradigms in Supply Chain Management (LARG_SCM)*. Supply Chain Management, InTech, 2011.

[75] Manufacturing: Leveraging industry 4.0 to create smart factories, [online] Available: https://www.bcx.co.za/wp-content/uploads/2019/01/Manufacturing.pdf.

12 Deep Learning Approach in Malware Hunting

P. M. Kavitha and B. Muruganantham

SRM Institute of Science and Technology, Chennai, India

CONTENTS

DOI: 10.1201/9781003156123-12

12.1 INTRODUCTION

Identifying the malware with its functions and its potential impact on any malware is defined as malware analysis. The malware code can be different radically since it is essential to know that the functionalities of malware can be many. Malware can be of any form; for example, it can be a virus, worms, spyware and Trojan horses [1, 2]. When an infected device contacts the healthy device, the malware gathers information or knowledge about the healthy device [3, 4]. In recent days, malware has become a huge order. Cuckoo sandbox, Net Cat, Dynamic Toll and Resource Hacker are some tools, which are generally used to perform malware analysis [5, 6]. But, these tools perform more in a traditional way. Whereas with the growing malware rate, new tools from Machine Learning (ML) and Deep Learning (DL) are implemented through this process [7– 9]. Many algorithms like Support Vector Machine (SVM), random forest, Convolution Neural Network (CNN) are used in the predictions of malware. These algorithms are used with the existing algorithms to make detection easier. In the academic and industrial sectors, malware researchers perform malware analysis better to understand malware and the various new methods in progress.

12.1.1 DEEP LEARNING

One machine learning technique is DL, which teaches computers about human actions: learning by example. Thus DL, a subset of AI, knows well to detect malware. DL is the idea behind driverless cars. This is done by the identification of stop signals and distinguishing the pedestrian post from a lamp post [10–12]. Also, the control is fed on customer services available in mobiles and tablets, televisions and speakers. Thus, DL gets a lot of attention from many users and researchers lately and the results achieved are tremendously improving at a growing rate. The computers are trained based on the classification from the text or images, or audio. The DL models achieve the finest accuracy, even exceeding the human level at times. A huge dataset with the label in a neural network architecture having layers is used to train the model [13– 16]. The applications of DL can be seen in industries like automation to medical sectors. The hidden layers in the neural network are usually referred to as deep [17– 19]. But the traditional neural network contained only 2–3 hidden layers, while in a deep network, the hidden layer can be up to 150. Based on the labeled dataset, the networks in DL are trained and modeled by which the features are learned from the data without any manual feature extraction [20, 21].

DL, which is a specific type of AI, is a quickly developing innovation. A machine learning methodology uses the relevant features from the manually extracted images

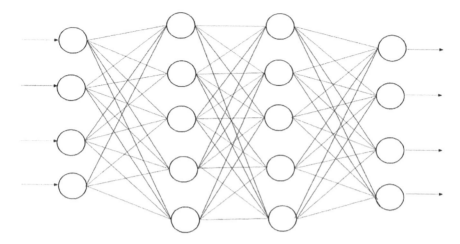

FIGURE 12.1 Neural network – deep learning structure [22].

[23, 24]. With these features, the models are created and categorized from the images. Figure 12.1 represents the traditional neural network structure. "End-to-end learning" is performed by the DL network, where the network gets raw data and performs a task like classification and gets to know about it [25–28].

12.1.2 ATTACKS

A noxious and intentional endeavor is either made by an individual or any association to break the data arrangement of another individual or association. Malware, phishing, Man-in-the-Middle assault, Denial-of-Service assault, SQL infusion, Zero-day misuse, DNS Tunneling are a portion of the normal sorts of accessible malware [29, 30].

12.2 LEARNING TECHNIQUES

Various methods of classification are used in the DL methodology [31–30]. The learning methods in DL are classified as "Supervised learning, Unsupervised learning, semi-supervised learning and reinforcement learning". The labeled structure of data is used in supervised learning. While in unsupervised learning, unstructured data is dealt with [34–38]. Under semi-administered learning, hardly any marked information and few unlabeled information are utilized [39–44]. In the reinforcement learning method, the output from the preceding layer is sent as the input for the current layer. CNN, ANN and RNN are some of the architectures in DL. Figure 12.2 represents the flow of control and work.

12.2.1 ARTIFICIAL NEURAL NETWORK

With Artificial Neural Networks (ANN), modeling complex ecosystems has become easy. They are used to predict the ecosystems and the changes they make in the environmental variables. Not only that, ANNs are used to identify the relationships

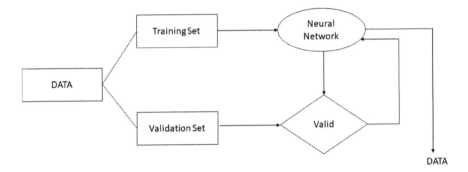

FIGURE 12.2 Deep neural architecture – flow of analysis using the training set and validation set.

among variables, making the ecosystem's function easier to understand. These calcu-
lations are utilized in taking care of immense issues to tackle a wide scope of issues,
according to Akmeliawati, Ooi and Kuang, 2007. In the first place, a variety of numbers is presented; xi is taken care of by the input layer of the handling hubs.
Intensification or hindrance is performed on the signs that move along the associations on each of the hubs in the neighboring layer through explicit weights, wt. For
the approaching signs, the adjoining hubs turn into the summation gadgets. At that
point, the yield signal (Oj) is recovered from the information signal, preparing units
after going through a limit work.

$$f(x) = 1i + e - x \tag{12.1}$$

In Equation 12.1, the range of the output $f(x)$ lies between 0 and 1, and the output
is calculated based on the following equation:

$$Q_j = 11 + e - \Sigma x x_i w_i. \tag{12.2}$$

The output signal Q_j is determined with the weighted associations with the accom-
panying layer of nodes. A similar cycle proceeds until the sign arrives at the yield
layer. The yield signal is then deciphered as the reaction of the ANN [45, 46].

12.2.2 CONVOLUTIONAL NEURAL NETWORK

The major building blocks in the convolution neural networks are the convolution
layers. Convolution characterizes as a straightforward use of channels to information
that brings about an actuation [47, 48]. The component map alludes to the rehashed
application on a similar channel to the information and results in a guide of actuation.
This demonstrates the areas and strength of the component in the info that is distinguished, for example, an image [49–51]. Figure 12.3 addresses the CNN structure.
This involves convolutional layers, tensor reshapes and a fully connected layer.

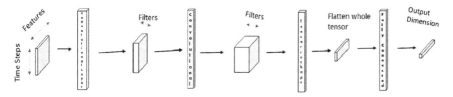

FIGURE 12.3   Convolution neural network [49].

12.2.3   RECURRENT NEURAL NETWORK

The data layer (input layer) gets the input, and the activations in the hidden layer are applied, then these activations are sent to the next hidden layer. Like this, subsequent activation is applied to the layers to produce the output [52, 53]. Each hidden layer has got its weight and biases. But, the weights and biases for the hidden layers are not the same, and hence, the values cannot be combined and are independent. Hidden layers with the same weights and biases may be combined. To do so, the recurrence formula is used on both e and h.

The formula for the existing state can be represented as $h_t = f(h_t - 1, x_t)$ [54]. Here in [54] ht is the representation of the new state; h_{t-1} represents the previous state; and x_t is the input given to the current state [55–58]. Having the previous state of input instead of the input itself makes the input neuron apply the previous inputs' transformations. Thus the successive input is known as the time step. Thus, four inputs may be obtained and given to the network during the recurrence formula. Considering the least complex type of recurrent neural network, the activation function is tanh, while W_{hh} is the weight at the recurrent neuron and Wxh is the weight of the input neuron. Thus, the equation for the state at time t is h$_t$ = tanh $(W_{HH}h_{t-1} + W_{hh}X_t)$ [59, 60]. The immediate previous state is taken into consideration by the recurrent neuron. Multiple states can be considered for longer sequences. With the final state calculated on hand, the output will be produced. The output is calculated based on $y_t = W_{hy} h_t$.

12.2.4   MULTILAYER PERCEPTRON NEURAL NETWORK

Cornel scientist Mr Frank Resenblatt proposed a simple perceptron based on the binary classification algorithm. Here, the arrangement of information images is separated into two sections: "yes" and "no" [57, 61, 62]. The perceptron model was arranged about the human psyche. The neuron has an uncanny limit for learning and dealing with multifaceted issues. This instrument has made the neural association not equivalent to other people. Not many information sources were taken, where each weight has a tremendous critical data which makes the yield decision as "0" or "1". Subsequently, uniting it with various other perceptrons, it outlines a fake neural association. With enough preparation and figuring power, a neural organization can respond to any question. A solitary perceptron collaborating with a few other perceptrons stacked in various layers takes care of the mind-boggling issues. The outline depicts the MLP having three layers [63, 64]. The initial layer on the left is the info layer that sends yields to all the perceptrons in the second layer (the covered-up layer).

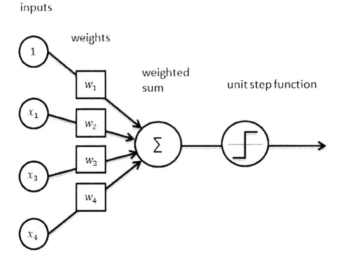

FIGURE 12.4   Perceptron structure [65].

The perceptron is the subsequent layer that presents the yield signs to the last layer on the privilege, the output layer. Figure 12.4 represents perceptron structure [65, 66].

12.2.5   GENERATIVE ADVERSARIAL NETWORK

GAN or the Generative Adversarial Network is used to generate and model the DL methods like CNN. This stands under the unsupervised learning method, which automatically discovers and learns the patterns and regularities in the input and gets modeled into output examples or samples plausibly derived from the original dataset. The two sections in a GAN are the generator and the discriminator [67–72]. The generator creates the conceivable information and the produced examples are shaped like the negative preparing set for the discriminator. At the same time, the discriminator distinguishes between the fake and the real data from the generator. Also, the penalty is imposed on the generator for the implausible results that are produced.

12.2.6   RESTRICTED BOLTZMANN MACHINE

The stochastic and generative Boltzmann Machines can learn the internal representations and represent them and also solve any difficult combinatoric problems. These machines named after Boltzmann distribution have become an integral part of statistical mechanics. These help to analyze the impact of temperature and entropy on the quantum states in thermodynamics.

12.2.7   DEEP BELIEF NETWORK

First-generation neural networks are implemented with perceptrons, and it is identified by specific objects based on the weight or the pre-fed properties. But, the perceptron works well only on the basic level and doesn't suit higher technology. Thus, the second generation of neural networks has taken its time to perform backpropagation. The output achieved will be compared with the expected output of zero error as a value. SVM are created to analyze the test cases by comparing them with the previous inputs. A directed acyclic graph, which is one form of a belief network, helps solve inference problems [8, 73, 74]. This is followed in the Deep Belief Networks, where unbiased values are stored in leaf nodes.

12.3 ACTIVATION FUNCTIONS

12.3.1 ROLE OF ACTIVATION FUNCTION

The inputs, which are the numeric data points in a neural network, are fed into the neuron's input layer. Figure 12.5 denotes the basic functionality of the activation function. The weight of each neuron is amassed with the value of the input that gives the output of the neuron, which is then moved on to the next layer [60, 75, 76]. The mathematical "gate" lies between the current neuron's input feed and the output that moves to the next layer. This simple step function is known as the activation function, where the neuron output is turned on and off, depending on the threshold. Likewise, a change can plan the information signals into the yield signals, which are fundamental for a neural organization. The neural organizations utilize the non-straight enactment work, which makes the organization learn unpredictable information, figure any capacities that address an inquiry and give an exact expectation.

12.3.2 TYPES OF ACTIVATION FUNCTION

12.3.2.1 Binary Step Function

The activation function based on the threshold is mentioned as a binary step function. When the threshold doesn't meet the threshold value, the neuron is actuated, and similar signs are transmitted off the following layer.

12.3.2.2 Linear Activation Function

The equation of a linear activation form can be given as $A = cx$ [77]. The inputs are accumulated with the neuron's weight, and an output signal is created, which is proportionate to the input. This demonstrates that a linear function is better than a step function as numerous inputs are allowed.

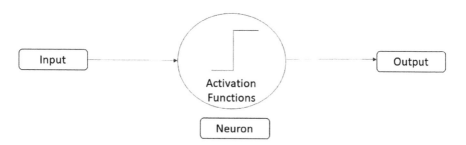

FIGURE 12.5   Activation function.

12.3.2.3 Non-Linear Activation Functions

Non-linear activation is utilized in present-day neural affiliation, which permits the model to make a fanciful orchestrating between the data and yield of the affiliation. Likewise, these are essential in learning and exhibiting the capricious data like pictures, video, sound, and instructive assortments of a non-direct set or one with high dimensionality. Any interaction can be addressed as a practical computational strategy in a neuron organization if the enactment work is straight.

12.3.2.4 Sigmoid/Logistic

The smooth gradient is the main feature in the activation function, which prevents "jumps" in the outputs. The output values are bound between "0 and 1", which standardize each neuron's outcome. The predictions are very clearly made for $-2 > X < 2$ and brings the Y value to the curve's edge, which lies close to 0 or 1. Thus clear predictions are made, but the activation of this activation function is a Vanishing gradient. There will be no change in predicting a vanishing gradient when the values are very high and low for X.

12.3.2.5 ReLU (Rectified Linear Unit)

The "ReLU" is mathematically effective as it permits the organization to connect quickly. Even though ReLU looks like a linear function, this is a non-linear function. It contains a derivative function that allows backpropagation. One such disadvantage of this function is the dying ReLU problem, where the gradient of the function turns to zero when the input reaches zero or has any negative value. In such a case, backpropagation cannot be performed by the network.

12.3.2.6 Leaky ReLU

This methodology prevents the dying ReLU by adding a small positive slope to the negative points and disabling the backpropagation for negative values. The results obtained will not be consistent as the leaky ReLU doesn't provide consistent predictions for negative inputs. The formula of leaky ReLU is

$$f(x) = \max(\alpha\partial, \partial). \tag{12.3}$$

12.3.2.7 Parameter ReLU

This allows a negative slope to learn, which is not like leaky ReLU and the negative part of the function is achieved through the slope, which is an argument. Thus, backpropagation is possible, and the required values of α are defined.

$$\sigma(z)j = \frac{ezj}{\sum_{k=1}^{k} EZK} \quad \text{for } j = 1\ldots k \tag{12.4}$$

12.3.2.8 Softmax

The Softmax can handle multiple classes, and only one class is in the activation function and it further normalizes the outputs on each class between 0 and 1. Then, the sum is divided and the probability of the input is delivered in the specific class. This stands to be useful for the output neurons, which means softmax is used for the output layer where the input is to be classified for the neural network in multiples.

12.3.2.9 Swish

A new "self-gated" activation function aims at better performance than ReLU, having a similar computational level in terms of efficiency. When experimenting with ImageNet on identical test cases by ReLU and Swish, swish tends to give top-1 classification accuracy of 0.6–0.9% higher than the other.

12.4 MALWARE ATTACKS

12.4.1 Introduction to Malware

The malware is malicious software, which means a "blanket of viruses, worms, Trojan and various other harmful programs", which is created to steal or destruct the data and access other systems' sensitive information [78–81]. A malware, in general, contains some suspicious code developed by the cyber attackers and is designed to damage or destruct the data and systems or gain authorized access to the systems. It is usually designed as a link or sometimes a file that can be mailed and when the user opens the link or the file, the malware begins its execution. Malware can use different ways to deliver its payload [48, 82].

12.4.2 Categories of Malware

One of the most common malware categories includes "virus", which executes itself and spreads and infects other programs and files. "Worms" are designed to "self-replicate" without any host and spreads without human interaction or directives from their malware creators. "Trojan horses" looks like a legitimate software program that gains access from a system. After the installation step, Trojan begins its malicious execution. "Spyware" is used to collect the information and data from the device and the user without their knowledge and observes their activities. "Ransomware" infects the system and encrypts the user's data [78, 83]. Then it's the turn of the cybercriminals to demand a lump-sum payment for the user to decrypt the system's data. A "Rootkit" is used to get the administrator-level of access on the victim's system. After the installation, the program threatens the user by getting privileged access to the system [84, 85]. A "backdoor" virus, known as Trojan (RAT), secretly creates a backdoor on the infected system and enables a threat and remotely access the system, which will not be known by the user or the security system programs [79, 86]. "Adware" tracks the user's browser and downloads the history and displays and pop-ups or advertises and makes sure to lure the customer's interests. For example, the advertiser uses cookies to track and track the web pages and advertise the recently visited pages. "Keyloggers", also known as system monitors, tracks every user's action on a computer [87–90]. This includes the emails, opened webpage, keystrokes and programs.

12.4.3 STAGES

The stages of a malware attack never remain the same on every attack, but it follows a similar attack. Thus, when the process strips up the hackers get huge money as always. And this states that the organization or user must have a great security feature to protect their information. Few steps are involved in investigating malware [91]. There are four stages in investigating malware, which forms a pyramid. As the pyramid moves forward, the complexity is increased. Moving from the bottom, the steps are discussed step-by-step. Fully automated analysis is the simplest way to access any suspicious program that is present. Scanning is done with fully automated tools. These tools give quick access to what the malware is capable of if it adulterates the system. A detailed report is produced on the network traffic, file activity and registry keys [34, 92]. This seems to be the fastest method to sift large malware in the system even if it doesn't provide a complete analysis.

12.4.3.1 Static Properties Analysis

To get a detailed look at the malware, an imperative analysis of the static properties is required. The properties are accessed more quickly as the malware is run, and it takes a lesser time. Hashes, embedded strings, embedded resources and header information are a few of the static properties. This property gives an elementary indicator [3, 93]. Interactive behavior analysis: An isolated library is generally required to observe a malicious file and see if it infects the laboratory [35, 94]. The analysts do a systematic analysis of the laboratory to check for malicious file entry on the hosts. Now, the analysts replicate the situation and identify the actions of the malicious file.

12.4.3.2 Manual Code Reversing

The code of the malignant record unravels the encoded information which is put away as the example and the rationale of the document's area is resolved to check for different capacities of that record, which don't appear on the social investigation. To physically invert the code, the malware examination devices like debugger and disassembler are utilized. These abilities to turn around the code are extremely fundamental, as it is minimal monotonous.

12.5 DEEP LEARNING OVER MALWARE ATTACKS

The process of determining the characteristics and intention of malware like viruses, worms, Trojans, etc. is known as malware analysis. The is a necessary step that is to be developed on the detection techniques on the malicious code. The two categories in the malware analysis tools are static analysis and dynamic analysis [95, 96]. Under the static tools, attempts to analyze the binary executables, without executing the value of binary. The live analysis or the dynamic analysis studies the binary's behavior after the execution. Automated malware analysis is more of a program that virtually interacts. But, it is not easy for a program to identify the exact behavior of another program.

12.5.1 Techniques

12.5.1.1 Signature-Based Detection

The signature-based approach or technique is primarily used on many antivirus software. Here, a unique signature is used from the captured malware file which is used to detect similar malware. And thus, a small false positive (FP) rate is produced. But, this doesn't seem to stop the attackers as they keep changing the malware signature that is detected by the antivirus [97]. This signature method is faster in detecting the malware, but it is difficult and fails in detecting new malware. This simply relies upon the execution of static investigation to extricate remarkable byte successions called marks.

12.5.1.2 Heuristic-Based Detection

Anomaly or behavior-based detection is also mentioned as Heuristic-Based Detection. Here, the malware's exercises at the run time are examined by a preparation stage. At that point, a mark is given as a noxious or genuine document while the testing or checking stage with the assistance of the example is separated during the preparation test [98]. Behavior-based detection helps in identifying both the unknown malware and the one that uses the obfuscation techniques. Excess monitoring time and a considerable false-positive (FP) rate are the drawbacks of this approach. In addition, the number of extracted features is reduced, and the similarities are evaluated between them, which monitors the malware activities that affect the ability to detect Zero-day malware attack. Data mining techniques are used here to understand the behavior of the file that is in progress, and the techniques like SVM, Naive Bayes, decision tree and random forest are used.

12.5.1.3 Static Analysis

The malicious software is analyzed before the execution state, which is known as static analysis. "String signature, byte-sequence-grams, syntactic library call, control flow graph and opcode frequency distribution" are the detection patterns that are used in static analysis. Unpacking and decryption are to be done on the executable file before the static analysis step. Various tools like disassembler or a debugger and a memory dumper can be used in reversing the compile windows executable file. IDA Pro and Olly Dbg are the disassembler or a debugger tool that helps display the malware code in Intel x89 assembly instructions. This provides a lot of insight into the malware's actions, and the patterns are provided to identify the attackers. Lord PE and Olly Dump are a few memory dumper tools that aim to extract the protected code located in the system's memory and dump it into a file. This useful technique is used in analyzing the executables that are packed and difficult to disassemble [99]. The twofold jumbling strategies are utilized in changing the malware parallels as self-compacted and organized paired records, which is exceptional and gets intended to oppose figuring out. This makes the static investigation both costly and untrustworthy. When the paired executables are used in a static investigation, different data like information designs or factor size will be lost and muddled. The avoidance strategies utilized by the malware programmers are advancing, and this prompts the improvement of dynamic investigation. The downsides of static investigation are

TABLE 12.1
Comparative Study of Static Analysis

Author and Year	Static Feature	Classifier	Dataset Malware/Benign	Acc	FP
Hashemi 2018	Opcode	KNN	M = 3,100 B = 3,100	91.9%	–
Salehi 2014	API, arguments	ROT-F, RF, DT, J48, NB	M = 826 B = 395	98.4%	3%
Han 2012	APIs sequence	–	M = 545	40%*	16%
Santos 2013	Opcode sequence	KNN, BN, SVM	M = 1,000 B = 1.000	97.5%	6%
Cheng 2017	Native APIs sequence	SVM	M = 18 B = 72	94.4%	1.4%

investigated by Moser et al. The work incorporates the presentation of a plan-based code muddling, which uncovers the reality that static examination alone isn't sufficient to distinguish the malware. Furthermore, the powerful investigation came right into it in this manner as it shows less weakness on code obscurity (Table 12.1).

12.5.1.4   Dynamic Analysis

Dynamic analysis refers to identifying and analyzing the behavior of the malicious code when it is still under execution state in a controlled environment. The controlled environments include a virtual machine, simulator, emulator, sandbox and few others. The perfect monitoring tools such as Process Monitor and Capture BAT, which are used for file system and registry monitoring, and Process Explorer and Process Hacker, which are used to monitor the process and replace Wireshark for monitoring the network and Regshot to change the system detection are used even before the malware samples are executed [98–100]. A few strategies are utilized to play out the unique investigation that significantly incorporates work call observing, data stream following, work boundary examination, guidance follows, and auto start extensibility focuses. In this manner, dynamic investigation is compelling than static examination which doesn't need to be dismantled. The malwares normal execution is a tough way in static analysis. However, it is a touch of tedious and asset devouring and permitting the adaptability issues to support up. There is significantly more contrast between the virtual climate and the genuinity of the malware in execution stage as the malware plays out a unique way of bringing about the counterfeit conduct rather than the genuine one. The malware conduct is likewise set off now and again on specific conditions that can't be recognized in the virtual climate. There are automated online tools such as Norman Sandbox, CW Sandbox, Anubis and TT Analyzer and ether and Threat Expert for the dynamic analysis of malware [101–105]. The analysis generated from these tools provides an in-depth knowledge of the malware and an insight into performed actions. An appropriate representation of malware is used in the classification of either similarity measure or feature vectors. An automated approach is required when many new malware samples arrive at the anti-vendors each day, limiting the samples. Machine

TABLE 12.2
Comparative Study of Dynamic Analysis

Author and Year	Dynamic Feature	Classifier	Dataset Malware/ Benign	Acc	FP
Liang 2016	API calls	DT, ANN, SVM	M = 12,199	91.3%	–
Mohaisen 2013	File system, registry, network	SVM, DT, KNN	M = 1,980	95%	5%
Mohaisen 2015	File system, registry, network	SVM, DT, KNN	M = 115,000	99%	–
Galal 2017	API's sequence	RF, SVM	M = 20,000/ B = 2,000	97.2%	–
Ki 2015	API's sequence	–	M = 23,080	99.8%	0%
Fan 2015	User API, native API	NB, SVM	M = 773/B = 253	95.9%	5%

learning techniques are used in the literature techniques for the automated malware analysis and classification (Table 12.2).

12.5.1.5 Hybrid Analysis

The hybrid analysis is used to gather information about the malware with the static and dynamic analysis. Also, this makes the scientific researchers get an advantage from both static and dynamic analysis. Thus, the ability of the malicious programs is detected in the right way. But the advantages and its limitations are on their side for both the analysis. Static analysis is cost-effective, faster and safer compared with dynamic analysis. But malware can make its way in any obfuscated method. Also, dynamic analysis stays reliable and controls and detects any obfuscation methods. And, it recognizes the malware variants and the other unknown malware families in a time-intensive and resource-consuming manner (Table 12.3).

12.6 RESULT

The MNIST dataset is used to calculate the malware classification. The dataset used herein generally contains 24594 malware under 2 classifications either in the malware or benign category. Also, from the stime and gtime the confusion matrix is calculated. The time taken for the malware to infect is determined with the stime and gtime factor. Thus, the malware is classified based on the classification and predicted based on the time stamp values. This given result states that the malware hunt based on the classification is possible using DL techniques (Figure 12.6).

12.7 CONCLUSION

This chapter is framed based on the study made on malwares and DL techniques. The dataset may either contain malware or a benign category, or even both. Here, the determination and identification of the same are done, which may help researchers

TABLE 12.3

Comparative Study of Hybrid Analysis

Author and Year	Feature Static/ Dynamic	Classifier	Dataset Malware/Benign	Acc	FP
Shijo 2016	PSI/API calls	RF, SVM	M = 1,368 B = 456	98.7%	–
Islam 2013	Function length, PSI/ API	DT, SVM, RF, IB1	M = 2,939	97%	5.1%
Ma 2016	Import functions/call functions	DT, NB, SVM	M = 279	–	–
Santos 2013	Opcode/system calls, operations, and exceptions	DT, KNN, NB, SVM	M = 13,189 B = 13,000	96.6%	3%

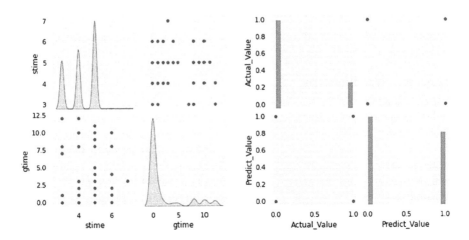

FIGURE 12.6 Confusion matrix calculated based on stime and gtime from the malware dataset.

work further on the dataset. With malware being detected, the time to infected time is calculated. With this, datasets having similar malware can be predicted for malware in the initial stage, and it helps for the accuracy and efficiency of the solution. This becomes one of the most acceptable methods to identify and predict malware based on its source of production and the infection time. As discussed above, malware analysis identifies and rectifies the malware and the issues caused by them. The future may combine two or three algorithms and deal with the malware, unlike using a single algorithm in this case. As a result of combining the algorithms or the techniques, efficiency and accuracy can be achieved at a higher rate. The supremacy of the two or more algorithms can be brought together in one frame and the malware can be dealt with.

REFERENCES

[1] M. Alazab, S. Venkataraman, and P. Watters, *"Towards understanding malware behaviour by the extraction of API calls,"* Proc. -2nd Cybercrime Trust. Comput. Work. CTC, vol. 2010, July 2009, pp. 52–59, 2010.

[2] Davis, A., Wolff, M., "Deep learning on disassembly data," April 2015, https://www.blackhat.com/docs/us-15/materials/us-15-Davis-Deep-Learning-OnDisassembly.pdf.

[3] M. Hafiz, M. Yusof, and M. R. Mokhtar," A review of predictive analytic applications of Bayesian network," *Int. J. Adv. Sci. Eng. Inf. Technol.*, vol. 6, no. 6, pp. 857–867, 2016.

[4] Z. Hu, W. Liu, J. Bian, X. Liu, and T.-Y. Liu, "Listening to chaotic whispers: A deep learning framework for news-oriented stock trend prediction," Dec. 2017, [Online] available: https://arxiv.org/abs/1712.02136

[5] M. Karresand, *"Separating Trojan horses, viruses, and worms—A proposed taxonomy of software weapons,"* in *IEEE Systems, Man and Cybernetics Society Information Assurance Workshop*, June 18–20, 2003, West Point, New York, USA. IEEE.

[6] I. Santos, F. Brezo, X. Ugarte-Pedrero, and P. G. Bringas, "Opcode sequences as representation of executables for data-mining-based unknown malware detection," *Inf. Sci.*, vol. 231, pp. 64–82, 2013.

[7] W. Wong and M. Stamp, "Hunting for metamorphic engines," *J. Comput. Virol.*, vol. 2, no. 3, pp. 211–229, 2006.

[8] The Microsoft cognitive toolkit, [Online] available: https://docs.microsoft.com/en-us/cognitive-toolkit/, 2017.

[9] A. Mohaisen, O. Alrawi, and M. Mohaisen, "AMAL: High-fidelity, behavior-based automated malware analysis and classification," *Comput. Secur.*, vol. 52, pp. 251–266, 2015.

[10] K. S. Han, I. K. Kim, and E. G. Im, "Malware classification methods using API sequence characteristics," *Lecture Notes Electric. Eng.*, vol. 120, pp. 613–626, 2012.

[11] A. Mohaisen and O. Alrawi, *"Unveiling Zeus: automated classification of malware samples,"* in *Proc. 22nd Int. Conf. World Wide Web companion*, Rio de Janeiro, Brazil, pp. 829–832, 2013

[12] U. Bayer et al., "Dynamic analysis of malicious code," *J. Comput. Virol.*, vol. 2, pp. 67–77, 2006.

[13] X. Ma, Q. Biao, W. Yang, and J. Jiang, *"Using multi-features to reduce false positive in malware classification,"* in *Proceedings of 2016 IEEE Information Technology, Networking, Electronic and Automation Control Conference, ITNEC 2016*, 2016, vol. 3, pp. 361–365.

[14] Adlice Software, "Rootkits hooks," 2014. [Online]. Available: https://www.adlice.com/.

[15] J. Berdajs and Z. Bosnic, "Extending applications using an advanced approach to DLL injection and API hooking," *Softw. Pract. Exp.*, vol. 40, no. 7, pp. 567–584, 2010.

[16] J. Butler, J. L. Undercoffer, and J. Pinkston, *"Hidden processes: The implication for intrusion detection,"* in *IEEE Systems, Man and Cybernetics Society Information Assurance Workshop*, USA, pp. 116–121, 2003.

[17] E. Gandotra, D. Bansal, and S. Sofat, "Malware analysis and classification: A survey," *J. Inf. Secur.*, vol. 5, no. 02, pp. 56–64, 2014.

[18] P. V. Shijo and A. Salim, "Integrated static and dynamic analysis for malware detection," *Procedia Computer Science*, vol. 46, pp. 804–811, 2015.

[19] Y. Huang, X. Ma, X. Fan, J. Liu, and W. Gong, *"When deep learning meets edge computing,"* in *Proc. IEEE 25th Int. Conf. Netw. Protocols (ICNP)*, pp. 1–2, Oct. 2017.

[20] Deep learning for Java: Open-source, distributed, Deep learning library for the JVM, 2017, [Online] available: https://deeplearning4j.org/

[21] Keras: The Python deep learning library, [Online] available: https://keras.io/

[22] F. Falah, O. Rahmati, M. Rostami, E. Ahmadisharaf, I. N. Daliakopoulos, H. R. Pourghasemi, "Artificial neural networks for flood susceptibility mapping in data-scarce urban areas," *Elsevier BV*, Germany, 2019.

[23] R. Islam, R. Tian, L. M. Batten, and S. Versteeg, "Classification of malware based on integrated static and dynamic features," *J. Netw. Comput. Appl.*, vol. 36, no. 2. pp. 646–656, 2013.

[24] X. Chen, J. Andersen, Z. Morley Mao, M. Bailey, and J. Nazario, *"Towards an understanding of anti-virtualization and anti-debugging behavior in modern malware,"* in *Proceedings of the International Conference on Dependable Systems and Networks*, Anchorage, pp. 177–186, 2008.

[25] H. Hashemi and A. Hamzeh, "Visual malware detection using local malicious pattern," *J. Comput. Virol. Hack. Tech.*, vol. 15, no. 1, pp. 1–14, 2018.

[26] A. Severyn and A. Moschitti, *"Learning to rank short text pairs with convolutional deep neural networks,"* in *Proc. 38th Int. ACM SIGIR Conf. Res. Develop. Inf. Retr. (SIGIR)*, New York, NY, USA, pp. 373–382, 2015.

[27] Z. Shao, L. Zhang, and L. Wang, "Stacked sparse autoencoder modeling using the synergy of airborne LiDAR and satellite optical and SAR data to map forest above-ground biomass," *IEEE J. Sel. Topics Appl. Earth Observ. Remote Sens.*, vol. 10, no. 12, pp. 5569–5582, Dec. 2017.

[28] N. S. Keskar, D. Mudigere, J. Nocedal, M. Smelyanskiy, and P. T. P. Tang, "On large-batch training for Deep Learning: Generalization gap and sharp minima," Sep. 2016, [Online] available: https://arxiv.org/abs/1609.04836

[29] S. Z. M. Shaid and M. A. Maarof, "Malware behaviour visualization," *J. Teknol.*, vol. 70, no. 5, pp. 25–33, 2014.

[30] Y. Cheng, W. Fan, W. Huang, and J. An, *"A shellcode detection method based on full native API sequence and support vector machine,"* in *IOP Conference Series: Materials Science and Engineering*, vol. 242, no. 1, pp. 1–7, 2017.

[31] G. A. N. Mohamed and N. B. Ithnin, "Survey on representation techniques for malware detection system," *Am. J. Appl. Sci.*, vol. 14, no. 11, pp. 1049–1069, 2017.

[32] Z. Bazrafshan, H. Hashemi, S. M. H. Fard, and A. Hamzeh, *"A survey on heuristic malware detection techniques,"* in *IKT 2013-2013 5th Conference on Information and Knowledge Technology*, pp. 113–120, 2013.

[33] S. N. Das, M. Mathew, and P. K. Vijayaraghavan, "An approach for optimal feature subset selection using a new term weighting scheme and mutual information," *Int. J. Adv. Sci. Eng. Inf. Technol.*, vol. 1, no. 3, pp. 273–278, 2011.

[34] M. Chowdhury and A. Rahman, *"Malware analysis and detection using data mining and machine learning classification,"* in *International Conference on Applications and Techniques in Cyber Security and Intelligence*, vol. 580, pp. 266–274, 2018.

[35] D. Ucci, L. Aniello, and R. Baldoni, "Survey on the usage of machine learning techniques for malware analysis," arXiv Prepr. arXiv1710, 08189, pp. 1–67, 2018.

[36] T. Teller and A. Hayon, "Enhancing automated malware analysis machines with memory analysis report," Black Hat USA, 2014.

[37] K.-W. P. Choi, Sang-Hoon, Yu-Seong Kim, *"Toward semantic gap-less memory dump for malware analysis,"* ICNGC Conf., Australia, pp. 1–4, 2016.

[38] R. Mosli, R. Li, B. Yuan, and Y. Pan, "A behavior-based approach for malware detection," *IFIP Adv. Informa. Commun. Technol.*, vol. 511, pp. 187–201, 2017.

[39] A. Damodaran, F. Di Troia, C. A. Visaggio, T. H. Austin, and M. Stamp, "A comparison of static, dynamic, and hybrid analysis for malware detection," *J. Comput. Virol. Hacking Tech.*, vol. 13, no. 1, pp. 1–12, 2017.

[40] M. Egele, T. Scholte, E. Kirda, and C. Kruegel, "A survey on automated dynamic mal-ware-analysis techniques and tools," *ACM Comput. Surv.*, vol. 44, no. 2, pp. 1–42, 2012.

[41] J. Korczak and M. Hemes, "*Deep learning for financial time series forecasting in A-Trader system*," in *Proc. Federated Conf. Comput. Sci. Inf. Syst. (FedCSIS)*, pp. 905–912, Sep. 2017.

[42] X. Yang, P. Zhao, X. Zhang, J. Lin, and W. Yu, "Toward a gaussian mixture model-based detection scheme against data integrity attacks in the smart grid," *IEEE Internet Things J.*, vol. 4, no. 1, pp. 147–161, Feb. 2017.

[43] X. Yuan, P. He, Q. Zhu, R. R. Bhat, and X. Li, "Adversarial examples: Attacks and defenses for Deep Learning," Dec. 2017, [Online] available: https://arxiv.org/abs/1712.07107

[44] Y. Yuan and K. Jia, "*A distributed anomaly detection method of operation energy consumption using smart meter data*," in *Proc. Int. Conf. Intell. Inf. Hiding Multimedia Signal Process. (IIH-MSP)*, Australia, pp. 310–313, Sep. 2015.

[45] X. Wang, W. Yu, A. Champion, X. Fu, and D. Xuan, "*Detecting worms via mining dynamic program execution*," in *Proceedings of the 3rd International Conference on Security and Privacy in Communication Networks, Secure Comm*, 2007, Nice, France, pp. 412–421.

[46] Z. Salehi, A. Sami, and M. Ghiasi, "Using feature generation from API calls for malware detection," *Comput. Fraud Secur.*, vol. 2014, no. 9, pp. 9–18, 2014.

[47] Q. Chen and R. A. Bridges, "Automated behavioral analysis of malware: A case study of wanna cry ransomware," arXiv Prepr. arXiv1709.08753, pp. 1–9, 2017.

[48] H. S. Galal, Y. B. Mahdy, and M. A. Atiea, "Behavior-based features model for malware detection," *J. Comput. Virol. Hacking Tech.*, vol. 12, no. 2, pp. 59–67, 2016.

[49] Z. Moti, S. Hashemi, and A. Namavar, "*Discovering future malware variants by generating new malware samples using generative adversarial network*," in *2019 9th International Conference on Computer and Knowledge Engineering*, Iran.

[50] Z. Zhao, J. Guo, E. Ding, Z. Zhu, and D. Zhao, "*Terminal replacement prediction based on deep belief networks*," in *Proc. Int. Conf. Netw. Inf. Syst. Comput.*, Wuhan, China, pp. 255–258, Jan. 2015.

[51] J. Zhu, Y. Song, D. Jiang, and H. Song, "A new deep-Q-learning-based transmission scheduling mechanism for the cognitive Internet of Things," *IEEE Internet Things J.*, doi: 10.1109/JIOT.2017.2759728.

[52] G. Willems, T. Holz, and F. Freiling, "Toward automated dynamic malware analysis using CWSandbox," *IEEE Secur. Privacy*, vol. 5, no. 2. pp. 32–39, 2007.

[53] A. Hosseini, "Ten process injection techniques: A technical survey of common and trending process injection techniques," 2017. [Online]. Available: https://www.end-game.com/blog/technical-blog/ten-process-injection-techniques-technical-survey-com-mon-and-trending-process, July 18 2017.

[54] AV-TEST, "The AV-TEST Security Report," Oslo, Norway, 2017. [Online]. Available at :https://www.av-test.org/fileadmin/pdf/security_report/AVTEST_Security_Report_2016-2017.pdf. April 1, 2018.

[55] Y. Ye, T. Li, D. Adjeroh, and S. S. Iyengar, "A survey on malware detection using data mining techniques," *ACM Comput. Surv.*, vol. 50, no. 3, pp. 1–40, 2017.

[56] K. Lee, H. Hwang, K. Kim, and B. N. Noh, "Robust bootstrapping memory analysis against anti-forensics," *Digit. Investig.*, vol. 18, pp. S23–S32, 2016.

[57] A. Moser, C. Kruegel, and E. Kirda, "*Limits of static analysis for malware detection*," in *Proceedings - Annual Computer Security Applications Conference, ACSAC*, Florida, pp. 421–430, 2007.

[58] J. Okolica and G. Peterson, *"A compiled memory analysis tool,"* in *IFIP Advances in Information and Communication Technology (AICT)*, vol. 337, pp. 195–204, 2010.

[59] P. Thomason, L. Wootton, E. Ford, and M. Nyflot, *"Deep convolutional triplet network for quantitative medical image analysis with comparative case study of gamma image classification,"* in *Proc. IEEE Int. Conf. Bioinformat. Biomed. (BIBM)*, USA, pp. 1119–1122, Nov. 2017.

[60] Y. Zhao, J. Li, and L. Yu, "A deep learning ensemble approach for crude oil price forecasting," *Energy Econ.*, vol. 66, p. 9–16, Aug. 2017.

[61] Ahmadi, M., Ulyanov, D., Semenov, S., Trofimov, M., and Giacinto, G., *"Novel feature extraction, selection and fusion for effective malware family classification,"* in *Proceedings of the Sixth ACM Conference on Data and Application Security and Privacy. CODASPY'16*, 2016, ACM, New York, NY, USA, pp. 183–194, https://doi.org/10.1145/2857705.2857713.

[62] C. Hong, J. Yu, R. Xie, and D. Tao, *"Weakly supervised hand pose recovery with domain adaptation by low-rank alignment,"* in *Proc. IEEE 16th Int. Conf. Data Mining Workshops (ICDMW)*, Spain, pp. 446–453, Dec. 2016.

[63] Endgame,"Ember," [Online]. Available: https://www.endgame.com/blog/technical-blog/introducing-ember-open-source-classifier-and-dataset, April 16 2018.

[64] Young, T., Hazarika, D., Poria, S., Cambria, E., "Recent trends in deep learning based natural language processing," CoRR abs/1708.02709, 2017, http://arxiv.org/abs/1708.02709, Aug 9 2017.

[65] E. Gandotra, D. Bansal, and S. Sofat, *"Integrated framework for classification of malwares,"* in *Proceedings of the 7th International Conference on Security of Information and Networks-SIN '14*, 2014.

[66] S. Gu, E. Holly, T. Lillicrap, and S. Levine, *"Deep reinforcement learning for robotic manipulation with asynchronous off-policy updates,"* in *Proc. IEEE Int. Conf. Robot. Autom. (ICRA)*,Singapore. pp. 3389–3396, May 2017.

[67] An open-source software library for machine intelligence, [Online] Available: https://www.tensorflow.org/, 2017.

[68] Theano, [Online] available: http://deeplearning.net/software/theano/, 2017.

[69] A. Adler, D. Boublil, M. Elad, and M. Zibulevsky, "A deep learning approach to block-based compressed sensing of images," Jun. 2016, [Online] available: https://arxiv.org/abs/1606.01519

[70] C. Angermueller, H. J. Lee, W. Reik, and O. Stegle, "DeepCpG: Accurate prediction of single-cell DNA methylation states using Deep Learning," *Genome Biol.*, vol. 18, no. 1, p. 67, 2017.

[71] C. Angermueller, T. Pärnamaa, L. Parts, and O. Stegle, "Deep Learning for computational biology," *Mol. Syst. Biol.*, vol. 12, no. 7, p. 878, 2016.

[72] I. J. Goodfellow et al., "Generative adversarial networks," ArXiv e-prints, Jun. 2014, [Online] available: http://adsabs.harvard.edu/abs/2014arXiv1406.2661G

[73] H. H. Huang and H. Liu, *"Big data machine learning and graph analytics: Current state and future challenges,"* in *Proc. IEEE Int. Conf. Big Data (Big Data)*, pp. 16–17, Oct. 2014; L. Huang, A. D. Joseph, B. Nelson, B. I. Rubinstein, and J. D. Tygar, *"Adversarial machine learning,"* in *Proc. 4th ACM Workshop Secur. Artif. Intell. (AISec)*, New York, NY, USA, pp. 43–58, 2011.

[74] E. Senft, P. Baxter, J. Kennedy, S. Lemaignan, and T. Belpaeme, "Supervised autonomy for online learning in human-robot interaction," *Pattern Recognit. Lett.*, vol. 99, pp. 77–86, Nov. 2017.

[75] P. V. Klaine, M. A. Imran, O. Onireti, and R. D. Souza, "A survey of machine learning techniques applied to self-organizing cellular networks," *IEEE Commun. Surveys Tuts.*, vol. 19, no. 4, pp. 2392–2431, 2017.

[76] K. Kowsari, D. E. Brown, M. Heidarysafa, K. J. Meimandi, M. S. Gerber, and L. E. Barnes, "HDLTex: Hierarchical deep learning for text classification," Sep. 2017, [Online] available: https://arxiv.org/abs/1709.08267

[77] C. T. Lin, N. J. Wang, H. Xiao, and C. Eckert, "Feature selection and extraction for malware classification," *J. Inf. Sci. Eng.*, vol. 31, no. 3, pp. 965–992, 2015.

[78] R. Mosli, R. Li, B. Yuan, and Y. Pan, "Automated malware detection using artifacts in forensic memory images," in *2016 IEEE Symposium on Technologies for Homeland Security, HST 2016*, 2016, pp. 1–6.

[79] A. Zaki and B. Humphrey, "Unveiling the kernel: Rootkit discovery using selective automated kernel memory differencing," *Virus Bull.*, no. September, pp. 239–256, 2014.

[80] M. H. Ligh, S. Adair, B. Hartstein, and M. Richard, *Malware Analyst's Cookbook and DVD: Tools and Techniques for Fighting Malicious Code*, Wiley Publishing, Boulevard, 2011.

[81] S. Kim, J. Park, K. Lee, I. You, and K. Yim, "A brief survey on rootkit techniques in malicious codes," *J. Internet Serv. Inf. Secur.*, vol. 3, no. 4, pp. 134–147, 2012.

[82] C.-I. Fan, H.-W. Hsiao, C.-Hs. Chou, and Y.-F. Tseng, "*Malware detection systems based on API log data mining*," in *2015 IEEE 39th Annual Computer Software and Applications Conference*, Taichung, Taiwan, pp. 255–260, 2015.

[83] S. T. Jones, A. C. Arpaci-Dusseau, and R. H. Arpaci-Dusseau, "*VMM-based hidden process detection and identification using Lycosid*," in *Proceedings of the fourth ACM SIGPLAN/SIGOPS International Conference on Virtual Execution Environments - VEE '08*, New York, pp. 91–100, 2008.

[84] K. T. Schütt, H. E. Sauceda, P.-J. Kindermans, A. Tkatchenko, and K.-R. Müller, "SchNet—A Deep Learning architecture for molecules and materials," *J. Chem. Phys.*, vol. 148, no. 24, p. 241722, 2018.

[85] D. Zhu, H. Jin, Y. Yang, D. Wu, and W. Chen, "*DeepFlow: Deep learningbased malware detection by mining Android application for abnormal usage of sensitive data*," in *Proc. IEEE Symp. Comput. Commun. (ISCC)*, Greece, pp. 438–443, Jul. 2017.

[86] A. Schuster, "Searching for processes and threads in Microsoft Windows memory dumps," *Digit. Investig.*, vol. 3, no. Suppl., pp. 10–16, 2006.

[87] D. Song et al., "*BitBlaze: A new approach to computer security via binary analysis*," in *Lecture Notes in Computer Science (including subseries Lecture Notes in Artificial Intelligence and Lecture Notes in Bioinformatics)*, vol. 5352 LNCS, pp. 1–25, 2008.

[88] M. Eskandari, Z. Khorshidpour, and S. Hashemi, "HDM-Analyser: A hybrid analysis approach based on data mining techniques for malware detection," *J. Comput. Virol. Hacking Tech.*, vol. 9, no. 2, pp. 77–93, 2013.

[89] I. Santos, J. Devesa, F. Brezo, J. Nieves, and P. G. Bringas, "OPEM: A static-dynamic approach for machine-learning-based malware detection," *Adv. Intell. Syst. Comput.*, vol. 189, pp. 271–280, 2013.

[90] C. Rathnayaka and A. Jamdagni, "*An efficient approach for advanced malware analysis using memory forensic technique*," *Proc. 16th IEEE Int. Conf. Trust. Secur. Priv. Comput. Commun. 11th IEEE Int. Conf. Big Data Sci. Eng. 14th IEEE Int. Conf. Embed. Softw. Syst.*, Sydney, pp. 1145–1150, 2017.

[91] N. Scaife, H. Carter, P. Traynor, and K. R. B. Butler, "*Crypto lock (and drop it): Stopping ransomware attacks on user data*," *Proceedings-International Conference on Distributed Computing Systems*, vol. 2016, pp. 303–312, August 2016.

[92] O. Z. Kraus et al., "Automated analysis of high-content microscopy data with Deep Learning," *Mole. Syst. Biol.*, vol. 13, no. 4, p. 924, 2017.

[93] A. Souri and R. Hosseini, "A state-of-the-art survey of malware detection approaches using data mining techniques," *Human-Centric Comput. Inform. Sci.*, vol. 8, no. 1. 2018.

[94] I. You and K. Yim, *"Malware obfuscation techniques: A brief survey,"* in *Proceedings – 2010 International Conference on Broadband, Wireless Computing Communication and Applications, BWCCA 2010*, Fukuoka, Fukuoka Prefecture Japan, pp. 297–300, 2010.

[95] D. Kirat and G. Vigna, *"MalGene,"* in *Proceedings of the 22nd ACM SIGSAC Conference on Computer and Communications Security - CCS '15*, Colorado, USA, pp. 769–780, 2015.

[96] V. Atluri an A. Chowdary, TRAN, "Botnets threat analysis and detection," Cham, 2017.

[97] P. T. Komiske, E. M. Metodiev, and M. D. Schwartz, "Deep Learning in color: Towards automated quark/gluon jet discrimination," *J. High Energy Phys.*, vol. 2017, no. 1, p. 110, Jan. 2017, [Online], available: https://doi.org/10.1007/JHEP01(2017)110

[98] T. Abou-Assaleh, N. Cercone, V. Keselj, and R. Sweidan, *"N-gram-based detection of new malicious code,"* in *Proc. 28th Annu. Int. Comput. Softw. Appl. Conf. 2004, COMPSAC 2004.* Hong Kong, vol. 2, no. 1, pp. 41–42, 2004.

[99] M. Egele, T. Scholte, E. Kirda, and C. Kruegel, "A survey on automated dynamic malware-analysis techniques and tools," *ACM Comput. Surveys*, vol. 44, no. 2, pp. 1–44, 2012.

[100] Y. Ki, E. Kim, and H. K. Kim, "A novel approach to detect malware based on API call sequence analysis," *Int. J. Distrib. Sens. Networks*, vol. 6, no. 659101, pp. 1–9, 2015.

[101] G. Liang, J. Pang, and C. Dai, "A behavior-based malware variant classification technique," *Int. J. Inf. Educ. Technol.*, vol. 6, pp. 291–295, 2016.

[102] Microsoft Azure, "What is a virtual machine?," 2018. [Online]. Available: https://azure. microsoft.com/en-in/overview/what-is-a-virtual-machine/.

[103] M. Sikorski and A. Honig, *Practical Malware Analysis: The Hands-On Guide to Dissecting Malicious Software.* E-book, no starch press. 2012.

[104] J. Stüttgen and M. Cohen, "Anti-forensic resilient memory acquisition, *Digital Invest.*, vol. 10, no. SUPPL., pp. 105–115, 2013.

[105] C. W. Tien, J. W. Liao, S. C. Chang, and S. Y. Kuo, *"Memory forensics using virtual machine introspection for Malware analysis,"* in *2017 IEEE Conference on Dependable and Secure Computing*, Taipei, Taiwan, pp. 518–519, 2017.

BLOG

[106] Network source, https://www.researchgate.net/figure/Deep-learning-CNN-model

[107] Image source, http://primo.ai/index.php?title=Perceptron_(P)

[108] https://www.codeproject.com/Articles/1206388/Build-Simple-AI-NET-Library-Part-Artificial-Neural

[109] https://www.analyticsvidhya.com/blog/2018/10/introduction-neural-networks-deep-learning/

13 Prediction of Breast and Lung Cancer, Comparative Review and Analysis Using Machine Learning Techniques

Arun Solanki, Sandeep Kumar, and C. Rohan
Gautam Buddha University, Greater Noida, India

Simar Preet Singh
Bennett University, Greater Noida, India

Akash Tayal
Indira Gandhi Delhi Technical University for Women, Delhi, India

CONTENTS

DOI: 10.1201/9781003156123-13

13.1 INTRODUCTION

Cancer is a general word that includes a diverse variety of disorders that can affect any portion of the body. Malignant tumors and neoplasms are other words used. One of the most distinguishing features of cancer is the accelerated formation of irregular cells that develop outside their usual borders and can then infiltrate adjacent areas of the body and migrate to other tissues, a metastatic phenomenon. Metastases are a leading cause of cancer-related mortality. Lung cancer is the most prevalent cancer in both men and women, and it is a considerable health problem worldwide. Cancer is the second leading cause of death worldwide, accounting for nearly 10 million deaths per year. Cancer is responsible for almost one in every six deaths worldwide [1]. Breast cancer grows in the breast cells and is the second most common cancer in women worldwide, after skin cancer. Breast cancer can affect both men and women, although it is far more common in women [2, 3].

It is necessary to manage this to prevent it from spreading to other areas of the body. Lack of physical exercise, tobacco use, a high BMI, alcohol use and a low fruit and vegetable intake are the causes of cancer and one-third of cancer deaths. In low- and middle-income countries, late-stage appearance and lack of access to care are normal. According to studies, intensive healthcare is available in more than 90% of high-income nations but less than 15% of low-income countries [4]. Cancer has a significant and rising economic influence. In 2010, the gross average financial burden of cancer was projected to be 1.16 trillion dollars. In 2019, just one out of every three countries reported high-quality cancer incidence statistics. The typical techniques for identifying cancer are C.T. scans and radiographs. These determine a modern and practical approach for identifying lung cancer at an early stage [4].

The most operative technique of finding lung nodules is C.T. scan. The use of a C.T. Image with computer processing to aid lung nodule diagnosis is widespread in clinics [5]. A screening system and a diagnostic system are two aspects of the computer-aided diagnosis (CAD) process for lung cancer. The CAD system aims to sort observed nodules into benign and malignant categories. Successful features such as texture, size, and growth rate can help CAD differentiate between benign and malignant pulmonary nodules. As a result, a CAD system's performance can be calculated using diagnostic accuracy, frequency, and automation level [6]. The majority of the proposed study is focused on deep Convolutional Neural Networks (CNN), with supervised learning being used in just a few cases. CNN's are a form of neural network that learns convolution constraints from a collection of available data during training. The authors are made up of multiple layers such as convolutional layers, de-convolutional layers, pooling layers, and so on [7].

This chapter discusses the more recent machine learning methods like Logistic Regression, SVM, Naïve Bayes, Decision Tree, Random Forest, and KNN for identifying lung cancer and breast cancer. In the literature review section, the past researcher`s methodology and dataset are discussed. In the Machine Learning techniques section, machine learning algorithms are defined with main features and formulas. The process flow diagram section defines the cancer detection model from the input dataset to performance analysis. The dataset and implementation section discusses dataset, dataset source and dataset features, and implementation of training, testing, and performance analysis with snapshots. This chapter's comparison and analysis section shows the performance of used algorithms with graphs and the conclusion section concludes the chapter in the conclusion section.

13.2   LITERATURE REVIEW

Support Vector Machine (SVM) to categorize the images into malignant or benign and scale it 1 to 5. The authors used SCM (Scale co-occurrence matrix) method for image feature extraction inside scale-space after wavelet transformation. It can help medical professionals in a more accurate diagnosis of lung nodule disease [8–10]. The authors proposed an approach based on a lung cancer dataset using the Artificial Neural Network (ANN) and found an accuracy of 93.33%. The researchers also mentioned that the combination of texture data and shape features data of the lung cancer dataset used for detection and classification would improve the classification accuracy [11]. The performance comparison between two different algorithms (KNN and Naïve Bayes) is compared and their accuracy is evaluated using the cross-validation technique. According to this chapter, the KNN gives the highest precision and lowest error rate than the Naïve Bayes classifier [11].

The authors compare different types of supervised ML Algorithms on Wisconsin Breast cancer Dataset (Original) with the algorithms: KNN, SVM, Naïve Bayes, and Decision Tree (C4.5). Their main aim was to determine each algorithm's classification accuracy, precision, sensitivity, and specificity in terms of efficiency and effectiveness. SVM has the highest accuracy (97.13%) and the lowest error rate. This study has used WEKA data-mining tools to run and simulate data [12, 13].

Past researchers have used different machine learning techniques and deep learning of Artificial Intelligence [14], which helps to apply in different fields and mainly in the medical field [15, 16]. In deep learning, CNN (convolutional neural network) helps image processing of the medical image to detect disease [17, 18].

13.3   MACHINE LEARNING TECHNIQUES

Machine Learning (ML) is the subfield of Artificial Intelligence. There are two types of ML algorithms based on input datasets: Supervised ML algorithms and Unsupervised ML algorithms [19, 20]. This study has used the supervised ML tools to analyze Lung Cancer and breast cancer using five different algorithms [21, 22]. The techniques are as follows:

- Logistic Regression classifier
- Support Vector Machine classifier
- Naïve Bayes classifier
- Decision Tree classifier
- K-Nearest Neighbor classifier

13.3.1   Logistic Regression Classifier

Logistic Regression is a classification method that can be extended to conditional classification problems, including the risk of contracting lung cancer (YES/NO). It employs the logistic or sigmoid function, which takes every actual value between 0 and 1 as input. The logistic function is used to estimate the likelihood of achieving something. The L.R. model always increases the value of variance if increased features are caused by the overfitting problem (Figure 13.1) [24].

13.3.2   Support Vector Machine (SVM) Classifier

SVM is used for two types of classification problems. The first one is linearly separable data and other non-linearly separable data using hyperplane with support

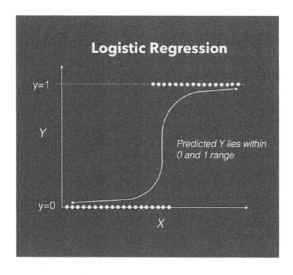

FIGURE 13.1   Logistic Regression representation graph [23].

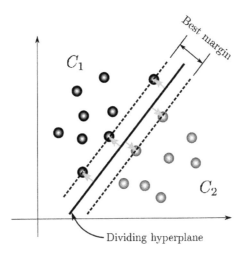

FIGURE 13.2   Support Vector Machine diagram [26].

vectors. Linear SVM is used for data divided into two groups using just a single straight line. For non-linearly separable data, non-linear SVM is used, which means if the dataset cannot be classified using a straight line, another dimension z needs to be added for this. That dimension is calculated as $z = x^2 + y^2$. After that, three-dimensional data converts into two-dimensional space again [25] (Figure 13.2).

13.3.3   NAIVE BAYES CLASSIFIER

This is also a supervised Machine Learning algorithm's technique that is established on the Bayes theorem, with robust expectations of independence among features. It is a simple technique for building the classifier assigned to the class labels function parameters drawn from the training set. Naïve Bayes adopts that every value of the attributes is inclusive of each other; that is, there are no possible associations between any two feature classes. The "Naïve Bayes algorithm is extremely accessible and performed in a complex situation very well. This is especially suitable when the dimensionality features have high inputs, which allow a huge number of linear function constraints" [27, 28].

This Naïve Bayes model technique is a conditional probability model and the conditional probability can be decomposed using Bayes' theorem as follows:

$$p(x) = \frac{p(Y_k)\,p(x|Y_k)}{p(x)} \qquad (13.1)$$

Here, Y_k is the dependent class variable and x is the independent variable and above conditional probability; the equation is also written below as

$$Posterior = \frac{Prior * Likelihood}{Evidence}$$
(13.2)

13.3.4   DECISION TREE CLASSIFIER

A Decision Tree looks like a tree structure in which internal nodes represent characteristics, branches represent decision rules, and leaf nodes represent outcomes. The Decision Tree aims to build an approach that forecasts the value of a target variable by analyzing a set of data and creating a set of laws used to forecast a class. The Decision Tree aims to generate a classifier method that predicts an output variable's value by examining data collection. In such a way, the Decision Tree chooses the best features for separating the records, converts those features into a decision node, divides the dataset into sub-sets, and then repeats these steps to construct the tree. Entropy and information gain help build the Decision Tree using variables (taken first) [29].

13.3.5   K-NEAREST NEIGHBOR CLASSIFIER

k-Nearest neighbors is also a supervised learning technique that stores the existing points and classifies new thresholds based on distance measurement. This methodology is non-parametric because no expectations about the distribution of the essential data are made, and it is also lazy because no training sample points model is created. In the evaluation process, all of the training data are included. It speeds up preparation while slowing down and increasing the expense of testing. If the number of groups is 2, the number of neighbors k usually is not an even number in this methodology. The distance between these data points is measured using the Euclidean distance method and Hamming distance to find the nearest related points [30, 31] (Figure 13.3).

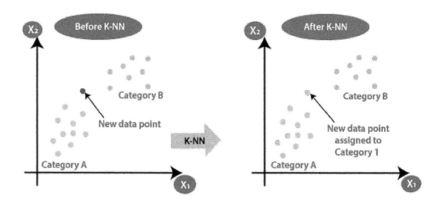

FIGURE 13.3   KNN classifier diagram [32].

13.4 PROCESS FLOW DIAGRAM

To analyze different researchers' approaches and decide how to improve the solution of the existing problem, cancer detection. The below process flow diagram is used for model training, testing, and performance analysis.

Step 1: Dataset Collection
This work collected the dataset from the secondary data sources for the model. In the next section, the dataset has been described in detail.

Step 2: Data Pre-Processing
Data pre-processing step is an essential part of model training. It helps remove a row or redundant data by applying pre-processing on the dataset to remove null values or fill missing values [33] (Figure 13.4).

Step 3: Feature selection
After cleaning the data by data pre-processing, select the best feature using the feature selection techniques to train with correct and desired features.

Step 4: Model training
Here, supervised ML algorithms trained on the input dataset to pass the selected feature data in the model training.

Step 5: Testing and Performance Analysis
In this step, a testing evaluation has been done. And finally, performance analysis is done and the result of the proposed model is also compared.

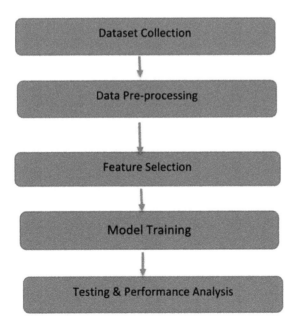

FIGURE 13.4 Flowchart of proposed system.

13.5 DATASET AND IMPLEMENTATION

In this chapter, different types of machine learning approaches are applied to detecting lung and breast cancer. The below table describes the analysis of five supervised machine learning approaches like Support Vector Machine (SVM), k-Nearest Neighbor (KNN), Decision Tree, and Naïve Bayes. These supervised learning algorithms apply on two different datasets: breast cancer dataset (that is, public dataset) and lung cancer dataset (available publicly). Our model gives additional accuracy on other datasets—the comparison of these model accuracies is provided in the table below. The dataset is divided into two parts to train and test the model. The proposed work used 25% of the dataset for testing and the rest 75% of the dataset for training purposes.

There are two datasets included in the chapter for cancer prediction:

- Breast cancer dataset
- Lung cancer dataset

13.5.1 BREAST CANCER DATASET

Breast cancer dataset downloaded from python's sklearn library as sklearn.datasets. **load_breast_cancer** [34] (Figure 13.5).

The breast cancer dataset is a binary classification dataset that is easy to use [30] (Figure 13.6).

	A	B	C	D	E	F	G	H	I	J	K	L	M	
1	id	diagnosis	radius_me	texture_n	perimeter	area_mea	smoothne	compactn	concavity	concave p	symmetry	fractal_di	radius_se	text
2	842302	M	17.99	10.38	122.8	1001	0.1184	0.2776	0.3001	0.1471	0.2419	0.07871	1.095	0
3	842517	M	20.57	17.77	132.9	1326	0.08474	0.07864	0.0869	0.07017	0.1812	0.05667	0.5435	0
4	84300903	M	19.69	21.25	130	1203	0.1096	0.1599	0.1974	0.1279	0.2069	0.05999	0.7456	0
5	84348301	M	11.42	20.38	77.58	386.1	0.1425	0.2839	0.2414	0.1052	0.2597	0.09744	0.4956	
6	84358402	M	20.29	14.34	135.1	1297	0.1003	0.1328	0.198	0.1043	0.1809	0.05883	0.7572	0
7	843786	M	12.45	15.7	82.57	477.1	0.1278	0.17	0.1578	0.08089	0.2087	0.07613	0.3345	0
8	844359	M	18.25	19.98	119.6	1040	0.09463	0.109	0.1127	0.074	0.1794	0.05742	0.4467	0
9	84458202	M	13.71	20.83	90.2	577.9	0.1189	0.1645	0.09366	0.05985	0.2196	0.07451	0.5835	
10	844981	M	13	21.82	87.5	519.8	0.1273	0.1932	0.1859	0.09353	0.235	0.07389	0.3063	
11	84501001	M	12.46	24.04	83.97	475.9	0.1186	0.2396	0.2273	0.08543	0.203	0.08243	0.2976	
12	845636	M	16.02	23.24	102.7	797.8	0.08206	0.06669	0.03299	0.03323	0.1528	0.05697	0.3795	
13	84610002	M	15.78	17.89	103.6	781	0.0971	0.1292	0.09954	0.06606	0.1842	0.06082	0.5058	0
14	846226	M	19.17	24.8	132.4	1123	0.0974	0.2458	0.2065	0.1118	0.2397	0.078	0.9555	
15	846381	M	15.85	23.95	103.7	782.7	0.08401	0.1002	0.09938	0.05364	0.1847	0.05338	0.4033	
16	84667401	M	13.73	22.61	93.6	578.3	0.1131	0.2293	0.2128	0.08025	0.2069	0.07682	0.2121	
17	84799002	M	14.54	27.54	96.73	658.8	0.1139	0.1595	0.1639	0.07364	0.2303	0.07077	0.37	
18	848406	M	14.68	20.13	94.74	684.5	0.09867	0.072	0.07395	0.05259	0.1586	0.05922	0.4727	

FIGURE 13.5 Breast cancer dataset divided into Malignant (M) and Benign (B).

Classes	2
Samples per class	212 (M),357 (B)
Samples total	569
Dimensionality	30
Features	real, positive

FIGURE 13.6   Breast cancer dataset details [35].

13.5.1.1   Breast Cancer Dataset Attributes

The main attribute of the dataset is described in Table 13.1.

13.5.1.2   Dataset Visualization of Breast Cancer

According to the target variable, using the python matplotlib library to visualize the system's dataset becomes more accessible to the readable and understandable form of Breast cancer with radius, texture, and perimeter mean (Figure 13.7).

The pair plot depicts the distribution of malignant and benign tumor data into two groups. In the pair plot, it's simple to differentiate.

13.5.1.3   Class Visualization of Breast Cancer

In a counterplot, the overall patient's counts with malignant (M) or benign (B) tumors are seen (Figure 13.8).

13.5.1.4   Correlation

The relationship between two variables is called their correlation. Here, the below correlation graph shows the correlation among the mean of breast cancer dataset features. This heat map shows how much one column stimulus all the other columns, like radius means, has 32% on texture mean (Figure 13.9).

TABLE 13.1
Dataset Attribute Details of Breast Cancer

S.N	Attributes
1.	I.D. number: This is like a primary key of the particular patient
2.	Diagnosis (M = malignant, B = benign): This is class/target variable
	Ten real-valued features are computed for each cell nucleus:
3.	Radius: The average of the distances between the middle and the points on the perimeter
4.	Texture: This is a standard deviation of gray-scale values
5.	Perimeter: The value of cells
6.	Area: The value covered by particular portions
7.	Smoothness: Local variation in radius lengths
8.	Compactness: This value is calculated as perimeter^2 / area - 1
9.	Concavity: The severity of concave portions of the contour
10.	Concave points: The number of concave portions of the contour
11.	Symmetry: The curve that separates equally
12.	Fractal Dimension: This is shown by coastline_approximation - 1

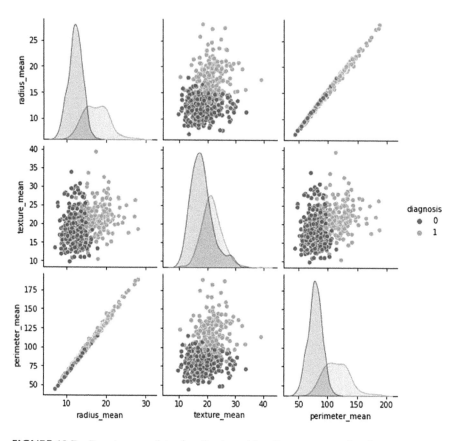

FIGURE 13.7 Breast cancer data visualization with radius, texture, and perimeter mean.

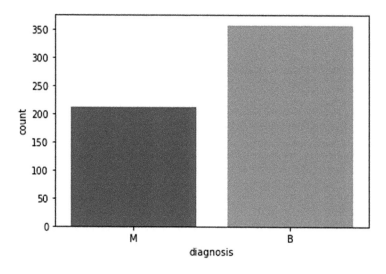

FIGURE 13.8 Class label of breast cancer.

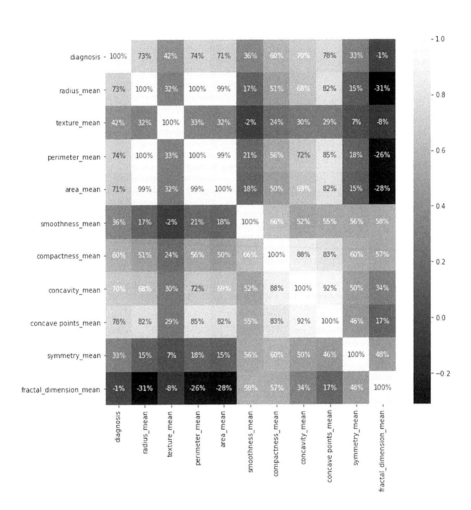

FIGURE 13.9   Correlation graph among means of breast cancer dataset features.

13.5.1.5 Implementation Snapshots (Figure 13.10)

```
In [22]: from sklearn.metrics import confusion_matrix

         for i in range( len(model) ):
           print('Model', i)
           cm = confusion_matrix(Y_test, model[i].predict(X_test))
           TP = cm[0][0]
           TN = cm[1][1]
           FN = cm[1][0]
           FP = cm[0][1]

           print(cm)
           print('Testing Accuracy= ', (TP + TN)/ (TP + TN + FN + FP))
           print()
```

```
Model 0
[[86  4]
 [ 3 50]]
Testing Accuracy=  0.951048951048951

Model 1
[[83  7]
 [ 2 51]]
Testing Accuracy=  0.9370629370629371

Model 2
[[87  3]
 [ 2 51]]
Testing Accuracy=  0.965034965034965
```

FIGURE 13.10. Testing accuracy of models - Model 0: Decision Tree, Model 1: Logistic Regression, Model 2: Random Forest.

13.5.1.6 Performance Metrics

Performance analysis is a specialized discipline that entails detailed analyses to improve efficiency and decision-making, mainly through quantitative statistical evidence. Accuracy metrics include precision, recall, f1-score, support values, and comparison with others proposed models.

Formulas for evaluation terms are as follows:

$$\text{Precision} = \text{True Positive} / \left(\text{True Positive} + \text{False Positive}\right) \tag{13.3}$$

$$\text{Recall} = \text{True Positive} / \left(\text{True Positive} + \text{False Negative}\right) \tag{13.4}$$

$$\text{F1 Score} = 2 * \left(\text{Recall} * \text{Precision}\right) / \left(\text{Recall} + \text{Precision}\right) \tag{13.5}$$

13.5.1.7 Accuracy Metrics (Table 13.2)

TABLE 13.2
Different Model Accuracy Metrics

	Precision	Recall	F1-score	Support
	Model (Decision Tree)			
'0'	97%	96%	96%	90
'1'	93%	94%	93%	53
Accuracy			95.1048%	
	Model (Logistic Regression)			
'0'	98%	92%	95%	90
'1'	88%	96%	62%	53
Accuracy			93.7062%	
	Model (Random Forest)			
'0'	98%	97%	97%	90
'1'	94%	96%	95%	53
Accuracy			96.5034%	

13.5.1.8 Accuracy Table

Table 13.3 shows that Random Forest performance is the best among the techniques.

TABLE 13.3
Accuracy Table of Different Techniques with Breast Cancer Dataset

S.N.	Technique	Dataset	Training Accuracy	Test Accuracy
1.	Decision Tree	Breast Cancer (569 Samples)	100%	95.10%
2.	Logistic Regression	Breast Cancer (569 Samples)	99.06%	93.70%
3.	Random Forest	Breast Cancer (569 Samples)	99.53%	96.50%

13.5.2 Lung Cancer

Lung cancer dataset is downloaded from kaggle [36]. The dataset description is below with several features and instances [37, 38] (Figure 13.11).

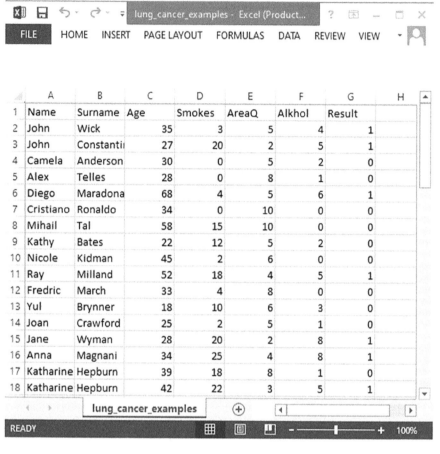

FIGURE 13.11   Lung cancer dataset with two classes of cancer: predicted (1) and non-predicted (0).

13.5.2.1   Attribute Information

The "Result" attribute is the class label in the form of 0 and 1.

'0' for no lung cancer detected

'1' for lung cancer detected.

A total of 59 rows and 7 columns are there in this lung cancer dataset. All predictive attributes are nominal, taking on integer values.

Age, Smokes, AreaQ, Alkohol

13.5.2.2   Dataset Visualization of Lung Cancer

The python matplotlib library visualizes the system's dataset to become more accessible to the readable and understandable form of lung cancer with Age, Smoke, Area, and Alcohol (Figure 13.12).

13.5.2.3   Class Label Visualization

In a counterplot, the overall patient's counts with malignant (M or 1) or benign (B or 0) tumors are seen respectively to result in '1' and '0' (Figure 13.13).

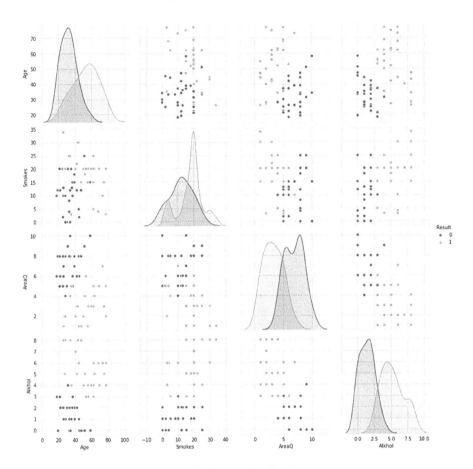

FIGURE 13.12 Lung cancer dataset's feature visualization.

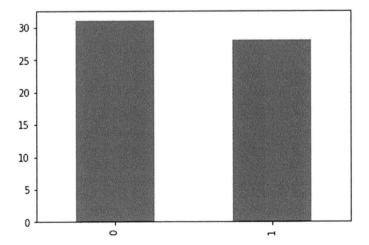

FIGURE 13.13 Class data visualization of lung cancer.

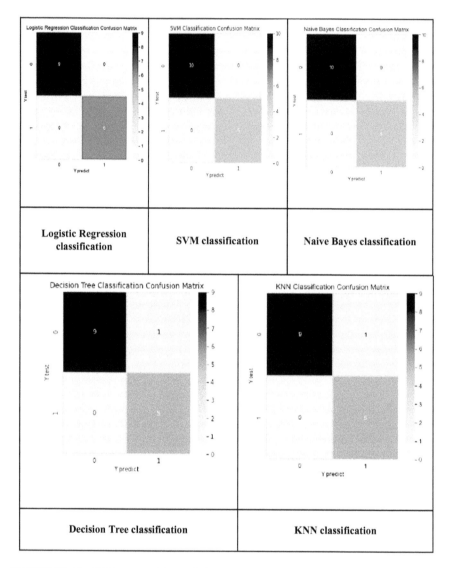

FIGURE 13.14   Different models confusion matrix.

Figure 13.14 shows that different models confuse matrix Model 1 (Logistic Regression classification), Model 2 (SVM classification), Model 3 (Naïve Bayes classification), Model 4 (Decision Tree classification), Model 5 (KNN classification).

13.5.2.4   Accuracy Table

Table 13.4 shows that KNN technique performance is the worst among the other techniques.

TABLE 13.4

Accuracy Table of Different Techniques with Lung Cancer Dataset

S.N.	Technique	Dataset	Accuracy
1	SVM	Lung Cancer (59 Samples)	100%
2	KNN	Lung Cancer (59 Samples)	93.0%
3	Decision Tree	Lung Cancer (59 Samples)	93.0%
4	Naïve Bayes	Lung Cancer (59 Samples)	100%
5	Logistic Regression	Lung Cancer (59 Samples)	100%

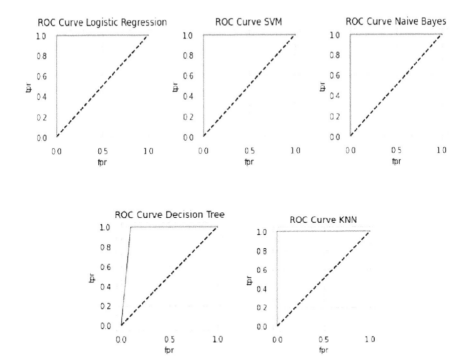

FIGURE 13.15   ROC curve on predicted scores.

13.5.2.5   ROC Curve

The ROC curve stands for Receiver Operating Characteristic curve; it is a graph that depicts a classification model's output on overall classification thresholds. There are

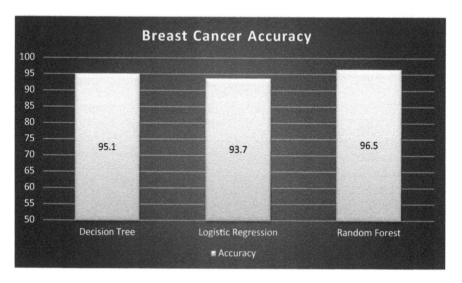

FIGURE 13.16   Comparison graph of breast cancer detection.

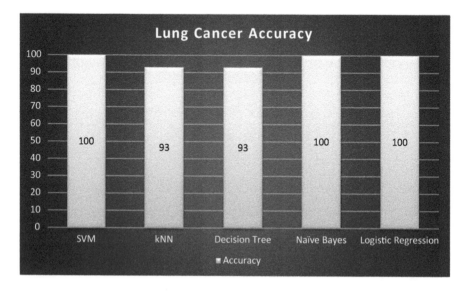

FIGURE 13.17   Comparison graph on lung cancer dataset.

two constraints on this curve: True Positive Rate measures how much something is true. The number of false positives is high (Figure 13.15).

13.6 COMPARISON AND ANALYSIS

The comparison between classification techniques on the performance of taken datasets (breast cancer and lung cancer) (Figures 13.16 and 13.17).

13.7 CONCLUSION

In this chapter, a textual dataset of cancer has been implemented for breast cancer and lung cancer using the supervised machine learning models. The implementation of the cancer detection task used python technology with popular python libraries such as numpy, pandas, matplotlib, and seaborn. The dataset splitting ratio is 25% for testing and 75% for training. This study compared five machine learning classifiers (SVM, KNN, Decision Tree, Naïve Bayes, and Logistic Regression) on lung cancer (textual dataset) and three machine learning classifiers (Decision Tree, Logistic, and Random Forest) on the breast cancer dataset. The experiment shows that Random Forest classifiers on the breast cancer dataset perform the best accurate results. SVM, Naïve Bayes, and Logistic Regression achieve the best result on the lung cancer dataset. The dataset is significant for building a solid machine learning model.

REFERENCES

[1] Cancer disease data, https://www.who.int/news-room/fact-sheet/detail/cancer, accessed on March 17 2021.
[2] Vaishnavi, D, Arya, K. S., Devi Abirami. T, Kavitha, M. N. 2019, Lung cancer detection using machine learning, *Int. J. Eng. Res. Technol.*, 7(01).
[3] Sadad, T., Munir, A., Saba, T., Hussain, A., 2018, Fuzzy C-means and region growing based classification of tumor from mammograms using hybrid texture feature, *J Comput. Sci.*, 29, 34–45.
[4] Awatramani, J. and Hasteer, N., 2019, *Early stage detection of malignant cells: A step towards better life*, International Conference on Computing, Communication, and Intelligent Systems (ICCCIS), Greater Noida, India, 2019, pp. 262–267, doi: 10.1109/ICCCIS48478.2019.8974543.
[5] Patra, R., 2020, *Prediction of lung cancer using machine learning classifier*, International Conference on Computing Science, Communication and Security, pp. 132–142.
[6] Chaudhary, A., Singh, S. S., 2012. *Lung cancer detection on CT images by using image processing*, International Conference on Computing Sciences.
[7] Kuruvilla, J. and Gunavathi, K., 2014, Lung cancer classification using neural networks for C.T. images, *Comput Methods Programs Biomed.*, 113(1), 202–209. doi: 10.1016/j.cmpb.2013.10.011. Epub 2013 Oct 18,. PMID: 24199657.
[8] Wu, Q. and Zhao, W., 2017, *Small-cell lung cancer detection using a supervised machine learning algorithm*, International Symposium on Computer Science and Intelligent Controls.
[9] Mohanambal, K., Nirosha, Y., Oliviya Roshini, E., Punitha, S., and Shamini, M., 2019, Lung cancer detection using machine learning techniques, *Int. J. Adv. Res. Electric. Electron. Instrum. Eng.*, 8(2), 266–271.

[10] Riquelme, D. and Akhloufi, M. A., 2020, Deep learning for lung cancer nodules detection and classification in C.T. scans, *A.I.*, 1(1),, 28–67. https://doi.org/10.3390/ai1010003

[11] Amrane, M., Oukid, S., Gagaoua, I. and Ensari, T., 2018, *Breast cancer classification using machine learning, Electric Electronics, Computer Science, Biomedical Engineerings' Meeting (EBBT)*, Istanbul, Turkey, 2018, pp. 1–4, doi: 10.1109/EBBT.2018.8391453.

[12] Asri, H., Mousannif, H., Al Moatassime, H., and Noel, T., 2016, Using machine learning algorithms for breast cancer risk prediction and diagnosis, *Proc. Comput. Sci.* 83, 1064–1069, ISSN 1877-0509, https://doi.org/10.1016/j.procs.2016.04.224.

[13] Chellamuthu, G., Kannimuthu, S., Premalatha, K., 2019, Chapter 11: Data mining and machine learning approaches in breast cancer biomedical research, *Sentiment Analysis and Knowledge Discovery in Contemporary Business*, IGI Global.

[14] Saba, T., 2020, Recent advancement in cancer detection using machine learning: Systematic survey of decades, comparisons and challenges, *J. Infect. Public Health*, 13(9), 1274–1289, ISSN 1876-0341, https://doi.org/10.1016/j.jiph.2020.06.033.

[15] Pandey, S. and Solanki, A., 2019, Music instrument recognition using deep convolutional neural networks, *Int. J. Inform. Technol.*, https://doi.org/10.1007/s41870-019-00285-y

[16] Rani, S. amd Solanki, A. (2021) Data imputation in wireless sensor network using deep learning techniques, Khanna A., Gupta D., Pólkowski Z., Bhattacharyya S., Castillo O. (eds), *Data Analytics and Management. Lecture Notes on Data Engineering and Communications Technologies*, Vol. 54. Springer, Singapore. https://doi.org/10.1007/978-981-15-8335-3_44

[17] Tayal, A., Gupta, J., Solanki, A. et al., 2021, DL-CNN-based approach with image processing techniques for diagnosis of retinal diseases, *Multimedia Syst.* https://doi.org/10.1007/s00530-021-00769-7

[18] https://www.cancer.org/cancer/cancer-basics/what-is-cancer.html, accessed on March 30 2021.

[19] Rajput, R. and Solanki, A., 2016, *Real time analysis of tweets using machine learning and semantic analysis, International Conference on Communication and Computing Systems (ICCCS-2016)*, Taylor & Francis, Dronacharya College of Engineering, Gurgaon, 9-11 September, pp. 687-692.

[20] Solanki, A. and Singh, T., 2021, COVID-19 epidemic analysis and prediction using machine learning algorithms, *Emerg. Technol. Battling Covid-19*, 324, 57, Nature Publishing Group.

[21] Levin, A. D., Ragazzi, A., Szot, S. L. and Ning, T., 2020, A machine learning approach to heart murmur detection and classification, *2020 13th International Congress on Image and Signal Processing, Bio Medical Engineering and Informatics (CISP-BMEI)*, Chengdu, China, pp. 521–525, doi: 10.1109/CISP-BMEI51763.2020.9263528.

[22] Tripathi, A. K., Garg, P., Tripathy, A., Vats, N., Gupta, D., and Khanna, A., Application of chicken swarm optimization in detection of cancer and virtual reality, Chapter-9 of *Advanced Computational Intelligence Techniques for Virtual Reality in Healthcare*, Springer (Scopus)

[23] Logistic regression image source, https://medium.com/@ODSC/logistic-regression-with-python-ede39f8573c7, accessed on April 30 2021.

[24] Maalouf, M., 2011, Logistic regression in data analysis: An overview, *Int. J. Data Anal. Tech. Strat.*, 3. 281–299. doi: 10.1504/IJDATS.2011.041335.

[25] Theodoros, E. and Massimiliano, P., 2001, Support vector machines: Theory and applications, 2049, 249–257,doi: 10.1007/3-540-44673-7_12.

[26] SVM image by Oscar Contreras Carrasco downloaded source, https://towardsdatascience.com/support-vector-machines-for-classification-fc7c1565e3 accessed on 1 May 2021.

[27] Chisaki, Y., Nakashima, H., Shiroshita, S., Usagawa, T., and Ebata, M., 2003, A pitch detection method based on continuous wavelet transform for harmonic signal, *Acoust. Sci. Technol.*, 24(1), 7–16.

[28] Solanki, A. and Saxena R., 2019, Text classification using self-structure extended multinomial Naive Bayes. *Handbook of Research on Emerging Trends and Applications of Machine Learning*, pp. 107–129.

[29] Patel, H. and Prajapati, P., 2018, Study and analysis of decision tree based classification algorithms, *Int. Jo. Comput. Sci. Eng.* 6, 74–78, doi: 10.26438/ijcse/v6i10.7478.

[30] Ahuja, R. and Solanki, A., *Movie recommender system using K-means clustering and K-nearest neighbor*, *Confluence-2019: 9th International Conference on Cloud Computing, Data Science & Engineering*, Amity University, Noida, doi: 10.1109/CONFLUENCE.2019.8776969

[31] Guo, G., Wang, H., David, B., Bi, Y., 2004, KNN model-based approach in classification.

[32] KNN algorithm image, https://www.javatpoint.com/k-nearest-neighbor-algorithm-for-machine-learning, accessed on May 7 2021.

[33] Data pre-processing using ML, https://www.javatpoint.com/data-preprocessing-machine-learning, accessed on May 1 2021.

[34] Breast cancer dataset source, https://scikit-learn.org/stable/modules/generated/sklearn.datasets.load_breast_cancer.html, accessed on April 15 2021.

[35] Breast Cancer dataset details image capture source, https://scikit-learn.org/stable/modules/generated/sklearn.datasets.load_breast_cancer.html, accessed on May 2 2021.

[36] Lung cancer dataset download source, https://www.kaggle.com/yusufdede/lung-cancer-dataset, accessed on April 28 2021.

[37] Lee, H. K., Ju, F., Osarogiagbon, R. U., Faris, N., Yu, X., Rugless, F., Jiang, S., and Li, J., 2017, A system-theoretic method for modeling, analysis, and improvement of lung cancer diagnosis-to-surgery process.

[38] Lin, D. and Yan, C., 2002, Lung nodules identification rules extraction with neural fuzzy network, *IEEE Neural Inform. Process.*, 4.0.

Index

Page numbers in *italics* denote figures.